AF766711

Reimagining Labor for a Sustainable Future

This book provides an original contribution to contemporary research surrounding the environmental, humanitarian and socio-political crises associated with contemporary capitalism.

Reimagining Labor for a Sustainable Future is guided by the assertion that new systems are always preceded by new ideas and that imagination and experimentation are central in this process. Given the vast terrain of capitalism—processes, institutions and stakeholders—Vogelaar and Dasgupta have selected labor as the point of engagement in the study of capitalist and alternative imaginaries. In order to demonstrate the importance of labor in rethinking and restructuring our world economy, the authors examine three diverse community projects in Scotland, India and the United States. They reveal the nuanced ways in which each community engages in commoning practices that re-center social reproduction and offer more expansive views of labor that challenge the neoliberal capitalist imaginary.

This book will be of great interest to students and scholars of sustainable economics, labor studies and sustainable development.

Alison E. Vogelaar is independent researcher and communication consultant. Her research and writing focus on the themes of representation, activism, labor and environmental communication. She has published numerous academic articles, one manuscript and two co-edited volumes. She was formerly an Associate Professor at Franklin University Switzerland (2008–2021) where she taught in the Communication and Media Studies program, founded and directed the university's SAFE Allies program and co-directed the Center for Sustainability Initiatives. She has recently joined the Academic Parity Movement as an advisor.

Poulomi Dasgupta is Associate Professor of Economics at Franklin University Switzerland, with research interests in political economy, development economics, postcolonial economics, labor and labor movements and sustainable development. She is also a research scholar at the Global Institute for Sustainable Prosperity. Her work on employment generation includes *The Job Guarantee: Toward Full Employment* and her work on sustainable development includes *Rethinking Economics: Experiences from Plural Socio-economic Higher.*

Routledge Studies in Sustainability

Smart Green World?
Making Digitalization Work for Sustainability
Steffen Lange and Tilman Santarius

Contesting Hydropower in the Brazilian Amazon
Ed Atkins

Circular Cities
A Revolution in Urban Sustainability
Jo Williams

Stories of Change and Sustainability in the Arctic Regions
The Interdependence of Local and Global
Edited by Rita Sørly, Tony Ghaye and Bård Kårtveit

Sustainable Living at the Centre for Alternative Technology
Radical Ideas and Practical Solutions
Stephen Jacobs

Rural Governance in the UK
Towards a Sustainable and Equitable Society
Edited by Adrienne Attorp, Sean Heron and Ruth McAreavey

The Environmental Impact of Cities
Death by Democracy and Capitalism
Fabricio Chicca, Brenda Vale and Robert Vale

Reimagining Labor for a Sustainable Future
Alison E. Vogelaar and Poulomi Dasgupta

For more information on this series, please visit: www.routledge.com/Routledge-Studies-in-Sustainability/book-series/RSSTY

Reimagining Labor for a Sustainable Future

Alison E. Vogelaar and Poulomi Dasgupta

Routledge
Taylor & Francis Group
LONDON AND NEW YORK

earthscan
from Routledge

First published 2023
by Routledge
4 Park Square, Milton Park, Abingdon, Oxon OX14 4RN

and by Routledge
605 Third Avenue, New York, NY 10158

Routledge is an imprint of the Taylor & Francis Group, an informa business

British Library Cataloguing-in-Publication Data
A catalogue record for this book is available from the British Library

ISBN: 978-0-367-68685-7 (hbk)
ISBN: 978-0-367-68686-4 (pbk)
ISBN: 978-1-003-13861-7 (ebk)

DOI: 10.4324/9781003138617

Typeset in Times New Roman
by Newgen Publishing UK

Contents

Acknowledgments

A book about the problematic nature of labor in neoliberal capitalism has been a funny thing to labor over for the past three years. Indeed, we have become strong advocates for the addition of a "who we wrote this book in spite of" section in books but alas that would take away from the people who deserve our pen and heartfelt thanks.

This begins with our editor, Annabelle Harris: Routledge is lucky to have you. So are we. Thank you for your patience, your kindness and for your trust in the project. We also could not have done this without the help of our editorial assistant, Jyotsna Gurung, and the project management support of Indhumathi Kuppusamy and editorial support of Ting Baker as well as the entire production team. Many thanks as well to the excellent feedback provided by the anonymous reviewers of our proposal.

We also want to thank the many collaborators unseen but central to this project. In particular, to Alexandra Peat, thank you for inspiring many parts of this book, for reading drafts, for listening to us, for checking in, for pushing us. To Iain Stewart, thank you for inspiring the chapter on Tombreck in Scotland and for being an incredible friend. To Shivani Shah, whose quiet encouragement played an important role in the completion of this project. To the friends who have supported us through this process, thank you.

In the spirit of the subject of this book, we would like to thank those beings who have reproduced us daily and yearly so that we could write a book about labor: to our parents, Len and Linda Vogelaar and Yashodhara and Sushanta Dasgupta: thank you for teaching us about the value of a job well done. To Mom (Linda), thank you for keeping us honest when we were supposed to be writing but you found us looking at dog photos on the internet instead.

To our partners, Ford Shanahan and Abhishek Mitra, thank you for the meals, the extra help, the taxi service, the childcare and your patience through the process. We know that this book about labor became your labor too!

To Neve and Elle, thank you for asking us so many good questions, most importantly, "is the book done yet?" and for being the reason we write about imagining better futures.

We would also like to thank Mitra TechLabs SP, Switzerland, for their generous help funding a research assistant position. We would also like to thank research assistant Thomas Liess for making our work smoother through his detailed, timely and consistent research support. Thank you as well to Callie Anderson for her early help researching cooperatives.

To our furry friends Stevie, Pixie, Banksy and Lucy thank you for your regular check-ins, tail wags and "requests" for food. Laboring alongside you is the only way we want to labor.

To the communities of Tombreck, Navdanya and Tonatierra, thank you for trusting us with this work and sharing your communities with us. A special thank you to Eve Reyes-Aguirres, Katy Macleod, Wendy Graham and Sue Manning for taking the time to help us understand the incredible work your communities do.

1 Introduction

Capitalism, crisis and the imagination

The public outrage over US presidential economic advisor Kevin Hassett's May 2020 declaration on national television that the nation's "human capital stock" was ready to return to work following COVID-related disruptions has proven something of a bellwether for the ways in which the pandemic changed and challenged work (Picchi, 2020). The Twitterstorm that ensued—with people like *Washington Post* syndicated columnist Catherine Rampell stating "that seems like a standard, pretty boring economic term to me" and *Economist* columnist Ryan Avent contending " 'human capital stock' is a stupid way to refer to people even if you are an economist talking to other economists about economics"—both laid bare festering tensions surrounding labor as imagined by the haves (the 1%) and the have-nots (the 99%) and reopened conversations about labor as a central but marginalized spoke in the capitalist wheel. In feminist economist and former New Zealand MP Marilyn Waring's terms, "human capital stock" represents one of the surfeit "clobber phrases" in economics; phrases that "hit you and leave you out cold" (2018, p. 22). Explaining their profusion in economics, Waring states: "I guess if you have always been a part of an ideology which colonizes words—labor, value and work, for example—you have never needed to evolve habits of clarity and specificity" (p. 571). This book is about these colonized words—labor, value and work—as they have been reanimated by the COVID pandemic.

Initially a public health crisis, COVID-19 "exposed and exacerbated a global crisis of productive and reproductive labor" (Stevano et al., 2021, p. 2). Early in the pandemic, lockdowns changed work patterns with some types of work temporarily relocated into the home (and often onto screens), others transformed to meet COVID needs and restrictions (e.g. wait staff doing delivery) and others (i.e. "essential workers") continuing to work in difficult and dangerous conditions. While no industry or workplace was left untouched by COVID-related disruptions, unemployment rates following the initial outbreak disproportionately affected laborers in the hospitality, retail and construction industries. In the longer term, COVID-related "restructuring" influenced many sectors and job types; however, according to OECD (2022), young, low educated, migrants and ethnic minorities were most affected across sectors. The ongoing nature of the pandemic triggered

DOI: 10.4324/9781003138617-1

conversations about the kinds of labor on which society depends but for which laborers are unrecognized and un- or underpaid (e.g. "essential work" and "care labor"). With many families juggling the extra burdens of care brought on by pandemic-related illness and the collapse of the home–work separation, this period saw the revitalization of discussions around care as an essential and unevenly distributed form of labor most often performed by women and marginalized members of society. While the "essential work" designation saw popular support for "frontline" workers in the form of global social-mediated-applauses, it saw little in the way of improvements to pay or work conditions. Workers noticed. In the context of the informal sector, the pandemic exacerbated the already precarious situation of workers who were forced to make life and death decisions to make ends meet (Mishra, 2020; Barhate et al., 2021). Over the course of two years, COVID-related disruptions to life and work resulted in what has been called the "Great Pause"—a moment of existential reflection, followed by the so-called "Great Resignation"[1]—a trend in parts of the Global North signaled by a subset of mid-career workers predominantly in the tech and health industries leaving stable employment because they were burned out.

We first conceived of this book in 2019 a full year before the pandemic and in the context of our unique biographical and disciplinary interests in the precarious plight of labor in the capitalist system and our shared experiences working in an increasingly neoliberal academic setting. We could not have foreseen how COVID-19 would influence our project and our lives. Even though our academic interests in labor were always anchored in our lived—classed, raced, gendered—experiences, this project hit very close to home. In addition to reinforcing the centrality of our topic, it also uprooted our own work-lives. We too found ourselves juggling the collapse of our home-work lives, overcome by the increased demands for care both at work and at home and caught in the financial maelstrom and organizational fallout that hit most universities. We adjusted our research schedules around childcare, quarantines and the loss of loved ones. Needless to say this book is "inspired by real events." We recognize however our relative privilege as knowledge workers living in a nation with social welfare programs and working in a sector that was, at least in the short term, insulated from the economic fallout of the pandemic.

Capitalism and crisis

As Stevano et al. importantly note, the "crisis of work caused by COVID-19 is not merely a tragic consequence of a freak epidemiological event, but rather a manifestation of the existing systemic fragilities of capitalism" (2021, p. 2). While crises have been the subject of socio-economic inquiry for at least 150 years, their contemporary resurgence reflects a heightened awareness of crisis phenomena since the 1970s (Jessop and Knio, 2018). The incredible global expansion of neoliberal capitalism in this period plays a central role. According to Bob Jessop (2013), crises are moments of contestation and struggle (over origins, causes and solutions). Because they "often produce profound cognitive, strategic and practical disorientation [crises] disrupt sedimented views of the world and disturb prevailing meta-narratives, theoretical frameworks, policy paradigms, and/or everyday life" (ibid., p. 68). In so doing they open a space for the proliferation of crisis interpretations, "only some of which get *selected* as the basis for imagined recoveries that are translated into economic strategies and policies—and, of these, only some prove effective and are *retained*" (ibid., p. 68). Because crises are never purely objective, material facts that imply a particular reaction, they "offer a real-time laboratory to study the dialectic of semiosis and materiality" as well as the dynamics of power on crisis construal, resolution and learning (ibid., p. 71). While it is true that the trope of crisis often functions as a deferral to "expert knowledge, specialized intervention, blueprints for action crafted by professional insiders" (Haiven, 2014, p. 31) or worse as a means of exploitation and profit (Klein, 2007), crises also create space for critical reflection and change (see Demaria, Kallis and Bakker, 2019 and Varvarousis, 2019).

Early critics of capitalism (e.g. Karl Marx, Max Weber, Thorstein Veblen) were straightforward about its inherent crisis tendencies ("contradictions" in Marxian terminology). Indeed, Marx's political economy is founded upon the assertion that crisis is at the heart of the capitalist system and that its periodic recessions, economic collapses, stock market crashes and their related effects (unemployment, poverty, homelessness and suicide) "were not simply momentary systemic lapses but rather the norm" and that these negative effects could not be avoided but only displaced and/or deferred (Haiven and Khasnabish, 2014, p. 97). Furthermore, all attempts to displace and defer these contradictions merely lead to new crises (e.g. overproduction related to automation, high production costs associated with paying fair wages, worker revolts related to oppressive conditions). Until the 1970s, policy approaches to capitalism were (albeit conservative) attempts to temper its crisis tendencies in the form of compromises with the laboring class. Neoliberalism saw an abrupt and unapologetic end to these compromises.

According to Andrew Gamble, there is general agreement among experts that the last century has witnessed three major economic crises: the depression of the 1930s characterized by deflation, the end of the Bretton Woods system in 1971 characterized by stagflation and the 2007–2008 global financial crisis characterized by "a new form of stagnation, leading to the notion of secular stagnation" (2018, p. 76). COVID-19 may very well be the fourth. While there is consensus regarding the occurrence of economic crises in these periods, the causes, scopes, dimensions and consequences of each crisis are the subject of contestation. The 2007/8 collapse of the global financial system and its correlate crises re-opened debates about the global economic system, inviting questions about its actual (diverse and contradictory) histories, compositions, beneficiaries, victims and consequences as well as its central myths and myth-makers. The crisis prompted a reconsideration of the inevitability of unfettered capitalism as a fact of life (Castells, 2017) and called into question the "fantasy of prosperity-through-growth" (Varvarousis, 2019, p. 12).

The post 2007/8 period saw the revitalization of the global justice movement and a new breed of "public square" protests as well as the renewal of academic and popular interest in the subject. Contemporary research highlights the system's negative correlation with personal and social well-being (e.g. Barry, 2012; Jackson, 2009; Layard, 2005), failure to result in development in the Global South and related "uneconomic" effects in wealthy nations (e.g. Alexander, 2012, Daly, 1999), its degrading effects on the health and integrity of the Earth's ecosystems (e.g. Global Footprint Network, 2010; IPCC, 2021) and collateral role in diminishing their capacity to support life in the future (e.g. Rockström et al., 2009) and its contribution to the re/production of various forms of socio- and geo-political inequity, instability and injustice (e.g. Elsner, 2012; Foster and Clark, 2020; Freedom in the World, 2018; Marazzi, 2009; Mies and Shiva, 1993; Nixon, 2013; Salleh, 2017).

In light of the mounting evidence of the system's failures, and in recognition of the pressing need to simultaneously raise the material standards of living of the world's poorest people while at the same time reducing humanity's overall ecological footprint (Meadows et al., 2004), recent years have seen the advancement of various approaches to economic restructure and reform. Some approaches (e.g. "sustainable development" and "green growth") approach the crisis from within the reigning system, while others (e.g. "degrowth," "solidarity economics" and "community economies") reject the growth-based macroeconomic policies on which it was built (Alexander, 2012; Hamilton, 2003; Purdey, 2010). Writing in 2012, degrowth scholar and advocate Samuel Alexander asserted that while "degrowth" in rich nations is the most viable solution to the challenges we are presently facing, it is "effectively unthinkable in today's politico-economic climate" (2012, p. 357; see also Fournier, 2008; Latouche, 2009; Jackson, 2009). While the neoliberal paradigm was given "a hideous afterlife in the wake of the 2007/8 financial collapse" in the form of financial industry bailouts and austerity measures, the contemporary acceleration of social and environmental breakdown combined with the maturation of inquiry and activism has begun to poke gaping holes in the ideology of neoliberal capitalism and there is widespread recognition that "the classical orientation towards the annual growth of production and consumption of market-mediated goods and services is no longer considered adequate" (Haiven and Khasnabish, 2014, p. 3; Brand, 2018, p. 148).

According to Nancy Fraser (2016), Marxism's narrow conception of capitalism constitutes a major impediment to understanding its inherent crisis tendencies. She and others (see Federici, 2012; Mies, 1986; Wren and Waller, 2018) advocate for an expanded understanding of capitalism that encompasses both its official economy and its "non-economic" background conditions asserting that "such a view permits us to conceptualize, and to criticize, capitalism's full range of crisis tendencies, including those centred on social reproduction" (2016, par 5). As an economic system, capitalism's accepted measures of "use value" and "exchange value" have given primacy to the exploitation of industrial men's "productive" labor but, significantly, exclude (non-value) the "reproductive" labor and resources on which all production relies (see also Folbre, 2006; Fraser, 2016; Mies, 1986; Waring and Steinem, 1988). As a result, most women's standing in this system (whether in the Global North or South) "is one of inclusion/exclusion—structurally essential to capital yet ambiguously defined as not quite labour, a condition of production or a natural resource" (Salleh, 2017, p. 11). Postcolonial and feminist scholars also highlight capitalism's dependence upon, and naturalization of, the "ongoing and violent expropriation of women, indigenous peoples, nonhuman animals, and the biosphere" (Oksala, 2018, p. 222). As a system that peripheralizes and does not value reproductive resources or labor, capitalism is an inherently unsustainable and unjust system that has, after 50+ years of unchecked expansion and integration, paved the way for a series of interlocking crises that can only be remedied outside the system (Salleh, 2017).

Contemporary approaches to capitalism's interlocking crises increasingly move beyond the limitations of existing models of "crisis management" framing the crisis not as *in* but *of* capitalism. According to Jessop, crises *in* occur within the parameters of a given set of natural and social arrangements and are associated with routine forms of crisis-management that restore basic features through "internal adjustments and/or shift crisis effects into the future, elsewhere, or onto marginal and vulnerable groups" whereas crises *of* occur when standard approaches to crisis management cease to work and/or efforts to defer or displace crises encounter growing resistance (2013, p. 74). Though we will never know whether the post 2007/8 critiques would have continued to be, as Fred Block put it, "like little yipping dogs at the heels of the neoliberal colossus, which can kick them away and proceed uninterrupted," or whether the tide had finally turned, it is clear that COVID-19 has proven something of a tipping point (in Healey and Barish, 2019, p. 5). We think labor is one reason why.

Before we continue, we want to pause on the word "labor." As an economist–non-economist research duo working on the complicated, abstract, and yet very material category of human activity we have come to call labor, we have had many interesting conversations about what we mean when we say the words "labor" or "work." What for an economist was a given, for a rhetorician was a riddle. The word "labor" comes from the twelfth century Latin for "task" or "toil" and is associated with difficult/undesirable forms of work. It became associated with a "class of laborers" from the late eighteenth and early-nineteenth centuries when early theorists of capitalism including Adam Smith, Karl Marx and Friedrich Engels began writing about the emergent system. The usage of "labor" versus "work" in the contemporary literature varies widely across disciplines and geographical/cultural contexts. As such, our book uses both "work" and "labor." We typically use "labor" when discussing ideas for which it is common economic parlance (e.g. productive and reproductive labor, wage labor) and "work" when we are referring more generally to laboring activities or, as with labor, when it is a common way of speaking (e.g. "home-work balance," "knowledge work"). We make every effort to use the terms intentionally and also recognize that they carry historical and political baggage (especially labor).

Labor and the crisis of capitalism

Labor was always fraught in the capitalist system. Indeed, Karl Marx's foundational thinking about the emergent system centered on the exploitation and alienation of labor—what he referred to as the central contradiction of capitalism—which he argued would lead to recurring crises and inevitable collapse. Part of the problem with labor is its narrow conception in classical economic thinking. Whereas early frameworks (i.e. Karl Marx and Adam Smith) placed the source of economic value in labor (labor theory of value) thereby giving some, albeit limited, visibility to the value of (productive) labor, neoclassical economists removed labor from economic analysis altogether when they shifted, in the late nineteenth century, to the utility theory of value in an attempt to construct a more "scientific" framework to study the economy. Labor in this conceptualization became a mere factor of production, like land and capital that could be exchanged in the market. It is in this framework that labor becomes framed and understood as a "disutility" by economists (i.e. something people would not choose to do if given the choice). Framed as a disutility, the focus on labor in economics turned to wages as the chief means of coaxing laborers into the undesirable task of working. Economist and sociologist Thorstein Veblen rejected neoclassical perspectives on labor asserting that humans are, on the contrary, driven by the "instinct of workmanship" and that the real issue with work was the "predatory instinct" of the capitalist class whose exploitation and domination of work processes made work undesirable (1898). Veblen's ideas about work were not new, Marx's contemporary William Morris is well known for decrying the alienation associated with industrial labor and advocating for a return to craftmanship.

The period of mass production and standardization associated with "Fordism" and "Taylorism" saw the intensification of the division of labor (deskilling), a process that accentuated worker alienation and precarity on the one hand and created the need for a managerial class on the other (Pietrykowski, 2019). Economic historian and anthropologist, Karl Polanyi, problematized this period's narrow view of the economy and labor asserting that the abstraction of the market in mainstream economics had created "fictitious commodities" of land, labor and money. Drawing on Marx, Polanyi asserted that labor could not (and thus should not) be reduced to a commodity, for it, like land, was not produced for the market. Indeed, Polyani believed that the capitalist economic system as conceived and constructed was so socially divisive that it had no long-term future.

The stock market crash of 1929 that followed the "roaring twenties" ushered forth widespread unemployment and a related shift in economic thinking about state intervention in matters of employment and social welfare. The brand of economic thinking and policy associated with John Maynard Keynes constituted a return to demand-side economic thinking, inserting the state as mediator between "capital" and "labor" and providing the necessary space for labor unions to develop. While the International Labour Organization was established in 1919 and labor/trade unions had been organizing since the rise of the Industrial Revolution, it was not until the 1950–1960s that they gained widespread acceptance and popularity. The postwar period was characterized by the relative strength of the laboring class in the context of national reconstruction efforts in the West and the continuation of Keynesian approaches to social welfare.

Against this backdrop, European and American intellectuals Milton Friedman and Friedrich Hayek were actively building a brand of economic thinking critical of Keynesian approaches to capitalism and outwardly fearful of the ascendance of state-socialism. The gradual cultivation of this brand of thinking would come to have widespread implications on economic policies and thought, chief among them, labor. Indeed, one of the greatest "successes" of neoliberalism has been the marginalization of labor in public discourse and policy. The "trente glorious" (1945–1975) or as de-growther Serge Latouche (2009) reframes it, the "trente pitious," saw the rapid rise and entrenchment of a growth-obsessed economic framework that saw the temporary growth of economies in some parts of the world but also set into motion the reproductive (social and environmental) crises we face today. This period is characterized by the intensification of the profit motive and the consequent neocolonial global expansion of market relations. The resultant global division of labor abruptly reshuffled older work patterns in both Global North and South with corporations using the tools of offshoring on the one hand and outsourcing on the other—what Weil and Goldman (2016) call the "fissured workplace"—to reduce the costs associated with labor. Across the globe these transformations have resulted in increased worker precarity and the declining quality of work conditions and relations. The expanding literature on the so-called "informal economy,"[2] "gig economy,"[3] and "need economy"[4] point to a radical departure in recent years from "traditional" wage labor in capitalist economies and the correlate decline in the middle-class (Pressman, 2007). The rapid automation of production and rise of Artificial Intelligence in the mid-2000s (the so-called "Fourth Industrial Revolution"[5]) have played a significant part in the "gigification" of the economy in some parts of the world as technologies are increasingly oriented to corporate growth and profit over social welfare and sustainability. The global financial crisis of 2007/8 set off by the collapse of the global financial services firm Lehman Brothers abruptly halted credit-based spending and growth and saw widespread defaults on loans resulting in home loss as well as loss of jobs, especially in real estate and finance (Good and Mance, 2011). While no nation was unscathed, the crisis most significantly impacted the economically integrated nations of North America and Europe, as well as Japan, Brazil and Botswana. The post-2007/8 period saw mounting dissatisfaction with policy responses that amounted to increased austerity measures on the one hand and financial industry bailouts on the other. COVID-19 appeared in early 2020 against this backdrop with a world economy dependent on the global flow of people and goods, whole populations unprotected by social safety nets and growing disenchantment with an economic system that had proven unsustainable and unequal yet somehow unstoppable.

Crisis of imagination

One intuitive approach to the challenge of economic transformation surrounds the idea of social imaginaries—an idea first outlined by philosopher and economist Cornelius Castoriadis in *The Imaginary Institution of Society* (1975). According to Castoriadis every society is organized around particular social imaginaries rooted in particular social-historical contexts and built upon founding myths. In addition to being "the creative and symbolic dimension of the social world ... through which human beings create their ways of living together and their ways of representing their collective life," social imaginaries make "possible common practices and a widely shared sense of legitimacy" (Thompson, 1984, p. 6; Taylor, 2004, p. 23). Our project is informed by Max Haiven's (2014) articulation of the contemporary crisis of capitalism as a failure of the *imagination*. According to the author, capitalism relies upon several accepted "imaginaries": of people as isolated, competitive economic agents, of the natural world (including human bodies) as an exploitable and renewable commodity, and most insidiously, of capitalism as "the natural expression of human nature, or too powerful to be changed, or that no other system could ever be as desirable" (ibid., p. 7). That is, it relies on us imagining that there is no other imaginary.

According to feminist economic geographers Katherine Gibson and Julie Graham, until recently most critiques of, and alternatives to, the reigning economic regime have been consigned to the margins of popular, political and academic discourses (Gibson-Graham, 2006). What is more, their marginality has been framed as a natural consequence of their inferiority. Central to shifting understandings of the crisis of capitalism is a growing recognition of "the peculiar power" bestowed upon it "to regenerate itself, and even to subsume its moments of crisis as requirements of its continued growth and development"—a power that has been achieved discursively through narratives that confer a mythic integrity and inevitability upon the economic-system-turned-life form (Gibson-Graham, 2006, p. 257; Brown, 2015; Haiven and Khasnabish, 2014). As such, capitalism has been relatively immune to critique and radical reconceptualization because contemporary discourses have endowed it with an extraordinary transformative capacity, rendering it "an agent that makes history but is not correspondingly 'made'" (2006, p. 38).

While the capitalist imaginary has clearly had a stranglehold on how we imagine, express, act and reproduce what we value, contemporary scholars find hope in the rise of alternative imaginaries, practices and communities. As we have demonstrated above, labor is both central to capitalist production and peripheral in its (and our collective) imagination. While most people spend a great portion of their lives laboring for all sorts of purposes, the capitalist imagination has limited it to the act of production for the market and is wholly negligent of the communal sources of reproduction on which the system depends. Our site-based study is informed by Gibson-Graham's "diverse economies" approach, which recasts capitalist hegemony as a dominant discourse as opposed to a social structure (Gibson-Graham, 2001). In an effort to partake in broader projects to imagine economic alternatives, and to recenter labor in this process, this book uses the framework of social imaginaries to showcase three diverse experimental communities who are laboring otherwise.

The following two chapters set up the framework for our case studies. Chapter 2 provides an interdisciplinary literature review of neoliberalism that situates it as the most recent formulation of capitalism distinguished by its global, albeit variegated, scope and, its fundamentalist commitment to profit. The chapter surveys contemporary shifts in thinking about the economy generally, and neoliberalism specifically, as a social construct rather than an immutable system. The chapter concludes with a survey of the consequences of neoliberalism on labor and how we think about work.

Chapter 3 offers a review of the literature on social imaginaries from its initial formulation by philosopher and economist Cornelius Castoriadis in the 1970s to its contemporary reformulation by cultural critics Max Haiven and Alex Khasnabish as a form of emancipatory practice. The chapter connects social imaginary research with the growing research on commons and commoning—specifically Gibson-Graham's community economies research as it helps articulate the necessary and natural bridge between language and practice. The chapter concludes with a description of how we will bring these literatures to bare on our research sites.

Chapters 4 through 6 present three diverse case studies of community projects that are reimaging labor in the spirit of building more sustainable and ethical societies. Chapter 4 explores a rural regenerative farm in Highland Perthshire Scotland, Tombreck Farm, as an example of an economic community that is creatively conjuring the past in order to reconstruct other-than-capitalist subjectivities, relations and networks. The chapter presents Tombreck's unique approaches to community development and natural regeneration, as they make it an exemplar of small-scale, place-based, embodied and culturally responsive approaches to social and economic transformation. Chapter 5 examines Navdanya, an organization and movement in rural India started by scholar-activist Vandana Shiva, as an example of a reproductive economy. The chapter studies Navdanya as it responds to an Indian economy that has been plagued by an agrarian crisis since the colonial period. The chapter examines the language and practices used by this site to reimagine labor and its role in social reproduction. Chapter 6 presents Tonatierra, a community organization based in Arizona, USA, that is creatively and dynamically responding to the impacts of colonialism and neoliberal capitalism on and around the US–Mexican border. The chapter focuses on two aspects of the organization as they symbolize its approach to repairing and re-connecting north–south and productive–reproductive divides.

Notes

1 See Geisler 2022, Cohen, 2021, Cook, 2021 https://hbr.org/2021/09/who-is-driving-the-great-resignation for further discussion.
2 See Sanyal 2007, Bangasser 2000, Stuart, Samman and Hunt 2018.
3 See Malik, Visvizi and Skrzek-Lubasinka (2021).
4 See Sanyal 2007.
5 See Schwab 2016.

References

Alexander, S. (2012) Planned economic contraction: The emerging case for degrowth. *Environmental Politics*, 21(3), pp. 349–368. doi:10.1080/09644016.2012.671569

Bangasser, P.E. (2000) *The ILO and the informal sector: An institutional history*. Geneva: International Labour Organization.

Barhate, B. et al. (2021) Crisis within a crisis: Migrant workers' predicament during COVID-19 lockdown and the role of non-profit organizations in India. *Indian Journal of Human Development*, 15(1), pp. 151–164. doi:10.1177/0973703021997624

Barry, J. (2012) *The politics of actually existing unsustainability: Human flourishing in a climate changed, carbon constrained world*. Oxford: Oxford University Press.

Block, F. (2018) *Capitalism: The future of an illusion*. Berkeley–Los Angeles: UC Press.

Brand, U. (2018) Growth, power and domination. In Jacobsen, S.G. (ed.). *Climate justice and the economy: Social mobilization, knowledge and the political*. New York: Routledge, pp. 148–167.

Brown, W. (2015) *Undoing the demos: Neoliberalism's stealth revolution*. Cambridge: MIT Press.

Castells, M. (2017) *Another economy is possible: Culture and economy in a time of crisis*. Hoboken: Wiley.

Castoriadis, C. (1975) *The imaginary institution of society*. Cambridge: MIT Press.

Cohen, D. (2021) The antidote to the great resignation is you. *HumanityWorks*, September 22. Available at: https://humanityworks.com/the-antidote-to-the-great-resignation-is-you/

Cook, J. (2021) Who is driving the great resignation?. *Harvard Business Review*, September 15. Available at: https://hbr.org/2021/09/who-is-driving-the-great-resi gnation

Daly, H. (1999) Uneconomic growth in theory and fact. The first annual Feasta lecture, April 16. Available at: http://feasta.org/documents/feastareview/daly/htm.

Demaria, F., Kallis, G. and Bakker, K. (2019) Geographies of degrowth: Nowtopias, resurgences and the decolonization of imaginaries and places. *Environment and Planning E: Nature and Space*, 2(3), pp. 431–450.

Elsner, W. (2012) Financial capitalism: At odds with democracy. *Real-World Economics Review*, 62, pp. 132–159. Available at: www.paecon.net/PAEReview/issue62/Elsner62.pdf

Federici, S. (2012) *Revolution at point zero: Housework, reproduction, and feminist struggle*. Oakland: PM Press.

Folbre, N. (2006) Measuring care: Gender, empowerment and the care economy. *Journal of Human Development*, 7(2), pp. 183–199.

Folbre, N. (2014) The care economy in Africa: Subsistence production and unpaid care. *Journal of African Economies*, 23(suppl_1), pp. 128–156.

Fournier, V. (2008) Escaping from the economy: The politics of degrowth. *International Journal of Sociology and Social Policy*, 28(11/12) pp. 528–545.

Foster, J.B. and Clark, B. (2020) *Capitalism and the ecological rift: The robbery of nature*. New York: Monthly Review Press.

Fraser, N. (2016) Capitalism's crisis of care. *Dissent*, 63(4), pp. 30–37.

Freedom in the World (2018) Report: Democracy in crisis. Washington DC: Freedom House.

Gamble, A. (2018) The 2008 crisis and the resilience of the neo-liberal order. In *The Pedagogy of Economic, Political and Social Crises*. New York: Routledge, pp. 75–87.

Geisler, J. (2022) The great resignation: Reality or myth?. Healthcare Financial Management Association, September 16. Available at: www.hfma.org/topics/hfm/2021/october/the-great-resignation--reality-or-myth-.html

Gibson-Graham, J.K. (2001) Re/presenting class. In Resnick, S. and Wolff, R.D. (eds.), *Essays in postmodern Marxism*, Durham, NC, and London: Duke University Press, pp. 1–22.

Gibson-Graham, J.K. (2006) *The end of capitalism (as we knew it): A feminist critique of political economy*. Minneapolis: University of Minnesota Press.

Global Footprint Network, 2010. Available from: http://footprintnetwork.org/en/index.php

Goodman, J. and Mance, S. (2011) Employment loss and the 2007–09 recession: An overview. US Bureau of Labor Statistics: Monthly Labor Review, April. Available at: www.bls.gov/opub/mlr/2011/04/art1full.pdf

Haiven, M. (2014) *Crises of imagination, crises of power: Capitalism, creativity and the commons*. London: Zed Books.

Haiven, M. and Khasnabish, D.A. (2014) *The radical imagination: Social movement research in the age of austerity*. London: Bloomsbury Publishing.

Hamilton, C. (2003) *Growth fetish*. Crows Nest: Allen & Unwin.

Healey, R. and Barish, J. (2019) Beyond neoliberalism: a narrative approach. *Narrative Initiative*. Available at: https://narrativeinitiative.org/resource/beyond-neoliberalism-a-narrative-approach/

IPCC (2021) Climate Change 2021: The Physical Science Basis. Contribution of Working Group I to the Sixth Assessment Report of the Intergovernmental Panel on Climate Change [Masson-Delmotte, V., P. Zhai, A. Pirani, S.L. Connors, C. Péan, S. Berger, N. Caud, Y. Chen, L. Goldfarb, M.I. Gomis, M. Huang, K. Leitzell, E. Lonnoy, J.B.R. Matthews, T.K. Maycock, T. Waterfield, O. Yelekçi, R. Yu, and B. Zhou (eds.)]. Cambridge and New York: Cambridge University Press. doi:10.1017/9781009157896. Available at: www.ipcc.ch/report/ar6/wg1/

Jackson, T. (2009) *Prosperity Without Growth? Economics for a Finite Planet*. London: Earthscan/Sustainable Development Commission.

Jessop, B. (2013) Putting neoliberalism in its time and place: A response to the debate. *Social Anthropology*, 21(1), pp. 65–74. doi:10.1111/1469-8676.12003

Jessop, B. and Knio, K. (eds.) (2018) *The pedagogy of economic, political and social crises: Dynamics, construals and lessons*. New York: Routledge.

Klein, A. (2007) Judging as nudging: New governance approaches for the enforcement of constitutional social and economic rights. *Columbia Human Rights Law Review*, 39, pp. 351–422.

Layard, R. (2005) *Happiness: Lessons from a new science*. New York: Penguin Press.

Latouche, S. (2009) *Farewell to growth*. Hoboken: Wiley.

Marazzi, C. (2009) *The violence of financial capitalism*. Cambridge: MIT Press.

Malik, R., Visvizi, A. and Skrzek-Lubasińska, M. (2021) The gig economy: Current issues, the debate, and the new avenues of research. *Sustainability*, 13(9), pp. 1–20.

Meadows, D., Randers, J. and Meadows, D. (2004) *Limits to growth: The 30-year update*. London: Earthscan.

Mies, M. (1986). *Patriarchy and accumulation on a world scale: Women in the international division of labour*. London: Zed Books Ltd.

Mies, M and Shiva, V. (1993) *Ecofeminism*. London: Zed Books.

Mishra, D.K. et al. (2020) Surviving the pandemic: Ground reports from India's villages, *Development Research Institute Bhubaneswar*. Available at: https://doi.org/10.13140/RG.2.2.19923.40486

Nixon, R. (2013). *Slow violence and the environmentalism of the poor*. Cambridge: Harvard University Press.

OECD (2022) "The unequal impact of COVID-19: A spotlight on frontline workers, migrants and racial/ethnic minorities," March 17. Available at: www.oecd.org/coronavirus/policy-responses/the-unequal-impact-of-covid-19-a-spotlight-on-frontline-workers-migrants-and-racial-ethnic-minorities-f36e931e/

Oksala, J. (2018) Feminism, capitalism, and ecology. *Hypatia*, 33(2), pp. 216–234.

Picchi, A. (2020) Trump adviser says America's "human capital stock" ready to return to work, sparking anger. CBS News, May 26. Available at: www.cbsnews.com/news/human-capital-stock-kevin-hassett-trump-economic-advisor-back-to-work/

Pietrykowski, B. (2019) *Work*. Cambridge: Polity Press

Pressman, S. (2007) The decline of the middle class: An international perspective. *Journal of Economic Issues*, 41(1), pp. 181–200.

Purdey, S. (2010) *Economic growth, the environment and international relations: The growth paradigm*. New York: Routledge.

Rockström, J. et al. (2009) Planetary boundaries: Exploring the safe operating space for humanity. *Ecology and Society,* 14(2). March 4. Available at: www.ecologyandsociety.org/vol14/iss2/art32/

Salleh, A. (2017) *Ecofeminism as politics: Nature, Marx and the postmodern*. London: Zed Books Ltd.

Sanyal, K. (2007) *Rethinking capitalist development: Primitive accumulation, governmentality and post-colonial capitalism*. India: Routledge.

Schwab, K. (2016) *The fourth industrial revolution*. Sydney: Currency.

Stevano, S., Ali, R. and Jamieson, M. (2021) Essential work: Using a social reproduction lens to investigate the re-organisation of work during the COVID-19 pandemic. SOAS Department of Economics Working Paper No. 241, SOAS University of London.

Stuart, E., Samman, E. and Hunt, A. (2018) *Informal is the new normal: Improving the lives of workers at risk of being left behind*. London: Overseas Development Institute.

Taylor, C. (2004) *Modern social imaginaries*. New York: Planet Books.

Thompson, J. (1984) *Studies in the theory of ideology*. Berkeley–Los Angeles: UC Press.

Varvarousis, A. (2019) Crisis, liminality and the decolonization of the social imaginary. *Environment and Planning E: Nature and Space*, 2(3), pp. 493–512.

Veblen, T. (1898) The instinct of workmanship and the irksomeness of labor. *American Journal of Sociology*, 4(2), pp. 187–201.

Waring, M. and Steinem, G. (1988) *If women counted: A new feminist economics*. San Francisco: Harper & Row.

Waring, M. (2018) *Still counting: Wellbeing, women's work and policy-making*. Wellington: Bridget Williams Books.

Weil, D. and Goldman, T. (2016) Labor standards, the fissured workplace, and the on-demand economy. *Members-only Library*, 20(1–2).

Wrenn, M.V. and Waller, W. (2018) The pathology of care. *Œconomia*, 8(2). Available at: https://journals.openedition.org/oeconomia/3195#quotation

2 The neoliberal imaginary

> It was what mathematicians called an attractor and astronomers a black hole: a massive blob of thought around which economic policy views revolved. The financial crisis of 2007–2009 shook the blob.
>
> (James K. Galbraith, 2021)

British Prime Minister Margaret Thatcher's now infamous slogan, "There Is No Alternative" (TINA) captures the smugness with which proponents of a neoliberal global order attempted to close economic debate in the 1980s. Thatcher's "TINA" signaled not only that debate was over but that neoliberalism had come out on top because of its natural superiority as opposed to its willful orchestration by elite special interests. For the prophets of the emergent system "deregulation [was] good, if not God" and selfishness was re-spun into the highest virtue of a well-functioning society (Flanders, 2013).[1] For two decades, the rhetoric of "TINA" stifled debate in many parts of the world: Consumers were apparently dazzled into compliance by their "freedom" to buy; poverty (and disease and destruction) were framed as inevitable and unavoidable side effects of Progress and Development that could be alleviated through aid and philanthropy but not cured; in the imaginary battle between the economy and the environment, the economy "naturally" won; and when the system finally and fantastically failed, the Too Big To Fail were necessarily "bailed out" while the too-small-to-matter were left to fend for themselves. This sweeping assessment of the seductions of "TINA," the duplicity of its advocates, ignorance of its patrons and erasure of its victims is a generalization that highlights some features of the contemporary hi/story of neoliberal capitalism but also masks its discontinuities, variations and resistance. The story of the "triumph" of the neoliberal economic regime is in fact much more messy and uneven and, crucially, *still being written.*

American economist James K. Galbraith describes neoliberalism as a "brand of economic thinking [that has] held pride of place" for the past 40 years (2021, par 1). While the concept is most frequently associated with the policy initiatives of Margaret Thatcher, Ronald Reagan and Alan Greenspan in the 1980s, neoliberalism

DOI: 10.4324/9781003138617-2

> has a complex intellectual history rooted in the interwar years in Europe and the USA … [it] was first trialled as an economic programme and political project in its Ordoliberal form in postwar West Germany and in its neoliberal form in Chile (thanks to the influence of "los Chicago boys").
> (Jessop, 2013, p. 68)

The contemporary "neoliberal movement"[2] was set in motion in the postwar period and is associated with the Mont Pelerin Society,[3] a think tank formed in 1947 in Switzerland during (and in antagonism with) the drafting of the world trade charter under the auspices of the International Trade Organization.[4] The group's founding fathers[5]—prominent European and American economists and intellectuals including Friedrich Hayek[6] and Milton Friedman— declared that "the central values of civilization," namely human dignity and freedom, were under threat and that their salvaging could be achieved by "intellectual argument and the reassertion of valid ideals" (The Mont Pelerin Society, 1947). Though the group's statement of aims is decidedly obscure, these "valid ideals" seem to center on the so-called universal "moral standards" and "rules of law" associated with classical liberalism and are set in opposition to socialist "ideologies." Their "statement of aims" asserts that the institutions of "private property and the competitive market" were the antidote to the so-called civilizational threats (of state-socialism). Though the founders "[sought] to establish no meticulous and hampering orthodoxy," there is no underestimating "The Friends of F.A. Hayek's"[7] role in naturalizing neoliberal philosophies and their related socio-political truisms. Several members of the society went on to pivotal positions advising and drafting economic policy and mainstreaming neoliberal economic theories well into the late twentieth century.

According to Stedman Jones (2013), neoliberalism's "center of gravity" shifted in the postwar period from Western Europe to the United States—more specifically to the Chicago[8] and Virginia Schools—where, over the course of two decades, it was distilled into an unflinching political mantra with a narrow focus on market liberalization and related opposition to most forms of state intervention in the economy. In spite (and perhaps because[9]) of the successes of postwar Keynesianism and Bretton Woods-inspired reconstruction efforts—which saw two decades of high growth rates in the US, UK and Europe, the subsequent growth of the middle class (see Toniolo, 1998; Klein and Pettis, 2021; Patnaik and Patnaik, 2016) and rise of active and successful labor unions—advocates of the neoliberal paradigm continued to argue its case in the postwar period.

Various historical accounts of "neoliberalism" (e.g. Harvey, 2005; Jones, 2013; Kotz, 2015) credit its far from certain ascendance in the 1970s and 1980s to the combined effects of widespread openness among experts and policymakers to economic reform, a series of cascading economic crises, and crucially, the framing of these crises as evidence of Keynesianism's failures. Economic reform in the direction of neoliberalism was catapulted by the demise of the Bretton Woods Agreement in 1971. While the Agreement is generally believed to have been a success in its early years, increasing disquiet in Western Europe over its systemic inequities (international exchange was tied to the dollar) and the overvaluation of the dollar in the context of economic decline in the US, led some European nations to exchange their dollar reserves for gold. This, combined with increasing unemployment rates and inflation in the US, led President Richard Nixon to "temporarily" suspend the Agreement, as part of his "New Economic Policy" of 1971. The so-called "Nixon Shock" marks the official end of the Bretton Woods Agreement and is credited with bringing on the stagflation of the 1970s. During the same period, many developing nations were facing the combined consequences of Import Substitution Industrialization (ISI) measures (implemented to decrease dependence on developing nations), bloated bureaucracies and stagnation related to economic declines in the West. This eventually led to debt crises in many countries in Latin America.

This period is also marked by major social changes and unrest manifesting in various social movements that diversely drew attention to problems of inequity and unsustainability and called for radical reform. According to Haiven,

> the radical movements of the 1960s and 1970s represented a monumental threat to global capitalism in that they demanded not only a bigger piece of the pie but a whole new society, one that offered workers and others real solidarity, autonomy, democracy, fulfillment and creativity.
>
> (2014, p. 244)

These movements ran parallel with the rise of "actually existing socialism" in many parts of the world and widespread support for communism; indeed many people either loosely advocated for pro-social policies or overtly sympathized with socialist regimes.[10] Such sympathies eventually made many groups targets of their own states as anti-communist alarmism was mobilized by those in power to stifle criticism of neoliberal policies and frame all forms of social welfare as inherently dangerous. These "first advances of the neoliberal counteroffensive" saw the systematic dismantling of forms of collective power, institutions of social welfare and frameworks of democratic governance, and a related pivot in social movement strategy with so-called "new social movements" shifting focus from radical political reform and economic redistribution to "identity politics"[11] (Murillo and D'Atri, 2018). According to West and Brockington, this period saw the most vocal critic of capitalism, the environmental movement, take a "step back from a posture of radical critiques of corporations, states, capitalism, and the collusion between the three" and insert itself instead into the power struggles over environmental governance in the recently decolonized nations of the South (2012, p. 2). Max Haiven (2014) also attributes the proliferation of Nongovernmental

Organizations (NGOs) in this period with the rise of an aging set of New Left activists seeking stable employment.

By the early 1980s, enduring stagflation in developed nations and mounting debt crises in developing nations inspired policymakers around the world to break from the Keynesian demand-side approach that had formed the basis of economic policy in the postwar period. Key markers of this break include the 1979 decision by the chair of the Federal Reserves, Paul Volcker, to increase interest rates in the US in an effort to assuage mounting (and increasingly politically mobilized) fears of continued inflation. While Volcker's intervention did not solve the problem of inflation, supply-side and corporate-centric approaches to economic reform and intervention persevered undeterred. According to many scholars, this commitment was the result of corporate-backed campaigns to "construct consent" for economic policies friendly to management over labor (Harvey, 2005, p. 44). This period saw the explosion of think tanks (e.g. the Heritage Foundation) and professional associations (e.g. the National Association of Manufacturers) dedicated to producing research and messaging to "create a climate of opinion" sympathetic to corporations and consumerism (ibid., p. 40). According to Harvey (2005), this "climate" was also enabled by the co-option and conversion of powerful cultural organizations and institutions like churches (e.g. the Evangelical "prosperity gospel"), the media (via advertising and buyouts) and universities.

In addition to its basis in the economic principles of supply side economics, monetarism and macroeconomic models of human behavior rooted in "rational expectations" (Best, 2020), neoliberalism is rooted in (selective evocations of) the political ideology of classical liberalism, most notably its celebration of individual freedom and subsequent designation of neoliberalism as its "exclusive guarantor" (Harvey, 2005: 40). It is connected to "free market fundamentalism"—a twentieth-century spin on laissez-faire economics that placed free markets (as opposed to states) at the ideological and institutional center of the economy. Guided by the ideology of "free markets," advocates argued that their "freedom" could only be realized via "free trade" agreements, limited government intervention, the systematic dismantling of labor unions (thereby stifling wages), tax cuts for the rich and the deregulation of financial markets. Of course, the decentering of the state in the affairs of the economy under neoliberalism was more rhetorical than literal and while the state disengaged from its prior role in the provisioning of social welfare (e.g. education, health), it became actively engaged in policymaking that prioritized capital over labor (e.g. the overt support for maintaining patents and copyrights in the name of innovation in developed nations). According to some critics, in the absence of social welfare social welfare, the state now maintains its legitimacy by manufacturing external enemies like terrorism, in order to create an "other" to unite the country and bolster some industries (Klein, 2007; Loewenstein, 2017; Patnaik, 2007).

The so-called "neoliberal agenda" was finally and fantastically propelled into the public consciousness, and sedimented into policy institutions, by two of its most ardent supporters, President Ronald Reagan in the US and Prime Minister Margaret Thatcher in Great Britain. Both leaders came to power in the midst of the major economic upheavals of the 1970s and early 1980s and were actively involved in developing neoliberal policy agendas resulting in large tax cuts for corporations, rampant deregulation (especially of the banking and finance sectors) and eventually to the spectacular financialization of the economic system. What is more, both leaders actively and publicly suppressed labor union activity.[12] In addition to instituting radical reforms in government regulation and spending, the Reagan and Thatcher administrations naturalized "neoliberalism" as the only viable economic system even as they rarely spoke its name: for, "what greater power can there be than to operate namelessly?" (Monbiot, 2016).

Naming and framing neoliberalism

The first critical literatures donning neoliberalism's name surfaced when "an identifiable family of context-specific 'neoliberalisms' first began to take shape—as concerted state projects and (anti)social programs—in locations like Chile, New Zealand, the United Kingdom and the United States" (Peck et al., 2018, p. xxii). Early Latin American inquiry drew attention to the radical, extreme and unorthodox character of neoliberal reforms in the region, the relationship of dependence and subordination with core economies, the legacies of authoritarianism and colonialism in its various implementations, and the need for a "third way" between, as Hugo Chavez put it, "savage neoliberalism and communist statism" (Garretón, 2004; Munck, 2003; Rodríguez, 2021; Stokes, 2001). English-language literatures followed suit and by 1994—with the passage of NAFTA and the Zapatista uprising in Chiapas—the term entered global circulation (Peck et al., 2018). According to Jamie Peck, social scientific deployments of the term neoliberalism have been

> curiously belated—picking up only in the period since the late 1990s, in part through critiques of orthodox globalization narratives, the follies of deregulation, and the failures of structural adjustment, but also paralleling the rise of global justice movements.
>
> (2018, p. xxii)

Scholarship on neoliberalism has proliferated since this time (Springer, 2016a; Wilson, 2017). In spite of its status as "an academic catchphrase" (Boas and Gans-Morse, 2009) and "deeply entrenched and normalized policy paradigm-cum-ideological commonsense" (Peck et al., 2018, p. xxii) the term remains obscure and is often "undefined in empirical research, even by those who employ it as a key independent or dependent variable" (Boas and Gans-Morse, 2009, pp. 138–139). Peck notes a tendency in the literature to use the term to signify "things we don't like" (2018, p. xxvii). As Noel Castree puts it, "neoliberalism is very much a critic's term" (2010, p. 1727) deployed by outsiders and critics of neoliberalism, as opposed to its advocates, and thus serving as Bob Jessop states, "more as a socially constructed term of struggle that frames criticism and resistance than as a rigorously defined concept that can guide research" (2013, p. 66). The 2008 global financial crisis and its fallout have seen neoliberalism enter the popular consciousness so much so that even a (minority) group of senior economists at the International Monetary Fund (IMF) have publicly acknowledged the negative consequences of the "neoliberal" policies they formerly espoused,[13] while other organizations like the Adam Smith Institute and Trump -era subreddit turned non-profit, The Neoliberal Project, have come out as neoliberal enthusiasts (see also Peck et al., 2018).

Given its ubiquity across disciplines and in the investigation of diverse phenomena, neoliberalism is conceptualized in diverse and often incommensurate ways. Birch and Springer (2019) distinguish between three analytic approaches to neoliberalism studies—historical/Foucauldian, critical/Marxist and ideational—that diversely conceive of neoliberalism as a form of governmentality, geographical process, hegemonic project privileging capital over labor, political project of regulatory and institutional change, set of transformative political-economic ideas, international development policy paradigm, epistemic community or thought collective, ideology or normative doctrine and/or form of organizational governance that has colonized and transformed political and other heretofore non-economic spheres and activities.

While the term's conceptual diversity may be construed as productive, if unavoidable, some critics assert that its plurality has come at the price of "leaving us more confused than enlightened" (Birch and Springer, 2019, p. 468; see also Boas and Gans-Morse, 2009 and Venugopal, 2015). As Peck et al. state,

> an unfortunate by-product of the proliferation of the term across the social sciences has been that, in many cases, the conceptualization of its effects becomes so totalizing and monolithic that it has progressively been imbued with its own causal properties.
>
> (2018, p. xxvii)

"It becomes," as Phelan puts is, "the 'it' which does the explaining, rather than the political phenomenon that needs to be explained" (2007, p. 328). Approaches to corralling this "rascal concept" vary (Brenner et al., 2010, p. 184); whereas many scholars stress the need to theorize neoliberalism with specificity and care (e.g. Peck, 2013; Peck et al., 2018; Phelan, 2014; Springer, 2014; Cahill et al., 2018), some assert that we may have hit "peak neoliberalism" in terms of its analytic and practical utility (Springer, 2016b) and others go so far as to question whether there is anything particularly unique about neoliberalism that would necessitate distinguishing it analytically from capitalism. Kean Birch and Simon Springer's introduction to the 2019 special issue in *ephemera*, "Peak Neoliberalism" asserts that notwithstanding concerns surrounding the potential for anti-neoliberal arguments to be seen as accepting "capitalism lite [...] there might yet still be some good reasons to salvage neoliberalism—or rather the thrust of the critique against it—from the dustbin of our contemporary intellectual toolkit" (2019, p. 473).

De-centering neoliberalism

> [Neoliberalism] is the outcome of two decades of specific political choices made by the most powerful actors in the world system.
>
> (George, 2004, p. 14)

Most scholars agree that it is no longer useful to think about neoliberalism as a global and/or universal process (Wacquant, 2012). As Cornelissen (2019) succinctly puts it, neoliberalism means different things at different times and in different places. Peck et al. assert that the definitional problems associated with neoliberalism are embedded within its constitutive features, which are "characterized by ongoing and highly variegated processes of market construction, which in turn generate feedback loops and contradictions, prompting further responses by policy makers and elites, thus rendering neoliberalism effectively indeterminate as a political project" (2018, p. xxviii). In this vein, many contemporary theoretical approaches to neoliberalism focus on its "variegated" nature noting the diverse impacts of local, historical, ecological, cultural and political factors and postcolonial settings (Wacquant, 2012). Bob Jessop distinguishes between four types of neoliberalism that emerged from the post-World War II "crisis" of capitalist development: Atlantic Fordism in advanced capitalist economies, import-substitution industrialization in Latin America and sub-Saharan Africa, export-oriented growth in East Asia, and state socialism in the Soviet Bloc, China and Indo-China (2013, p. 73).

Peck et al. offer the working concept of "actually existing neoliberalism" as a concept that serves three analytic purposes: First, it draws attention to the

> discrepancies between neoliberalism as a tutelary theory and its evidently variegated practice, between the utopian ideology of the free-market counter-revolution and its earthly manifestations, and between programmatic ambition of this frontal discourse and its frustrated, compromised, crisis-prone and yet restlessly experimental form.
>
> (ibid., p. 13)

Second, it highlights the diverse ways in which neoliberal restructuring strategies interact with pre-existing and co-existing social formations. And third, it posits uneven spatial development as an essential characteristic of neoliberalism. The authors contend that neoliberalism's reproduction can only be understood via "the dynamic mapping of these mongrel formations and the connective relations between them" (ibid., p. 13).

Notions of neoliberalism as variegated are complemented by contemporary literatures framing neoliberal capitalism not as a system or set of policies but rather as a hegemonic social construct constituted by subjects, relationships, institutions, practices, languages and meanings that have become "naturalized" but are by no means "natural." According to J.K. Gibson-Graham, "the birth of the concept of Capitalism as we know it coincided in time with the birth of 'the economy' as an autonomous social sphere" (2006a, pp. 253–254). "The economy," they contend, "has been performed and made 'true', coming into existence as something widely acknowledged and socially consequential, something that participates in organizing life and things within and around it" (ibid., p. xxi). In this context, the past 50 years have been characterized by a general understanding of capitalism as the dominant, if not exclusive, economic system and the "ultimate container within which we live" (ibid., p. 260). While political and corporate elites are certainly complicit in such framing, its detractors (chiefly Marxists) have also participated in the production of a discourse of "Capitalism" that represents it "as unified, singular and total rather than as uncentered, dispersed, plural, and partial in relation to the economy and society as a whole" (ibid., p. 259). Such framing positions capitalism as necessarily hegemonic by virtue of its own characteristics and has, according to Jessop (2013), allowed the system to "'fail forward', as its diverse proponents and fellow-travellers draw on its unevenly developing crisis-tendencies, contradictions, and resistances to renew the broad project in changing conditions" (pp. 76–77).

Cultural critic Mark Fisher uses the term "capitalist realism" to describe the "pervasive atmosphere conditioning not only the production of culture but also the regulation of work and education, [that acts] as a kind of invisible barrier constraining thought and action" and makes it difficult, if not impossible, to imagine coherent alternatives (2009, p. 16). According to the author, capital realism has "successfully installed a 'business ontology' in which it is simply obvious that everything in society, including healthcare and education, should be run as a business" (ibid., p. 17). Indeed, it has been so successful that even oppositional discourses are quickly co-opted and commodified making resistance feel futile and leading to a widespread and pervasive sense of apathy. Haiven reminds us that "the perversion, mutation and distortion of our hopes and dreams" is not a mere side effect of the system, it is fundamental to its functioning (2014, p. 26). As David Graeber asserts, "hopelessness isn't natural. It needs to be produced" (2008, p. 1). Graeber credits the "construction of a vast bureaucratic apparatus for the creation and maintenance of hopelessness" (e.g. armies, prisons, police, security firms, police and military intelligence apparatus, and propaganda engines) since the 1980s for its role in creating "a pervasive climate of fear, jingoistic conformity, and simple despair that renders any thought of changing the world an idle fantasy" (ibid., p. 1).

According to Gibson-Graham, the widespread but "silent consensus" among publics and experts of capitalism as "something large that shows us to be small" is key to its endurance (2006a, p. 253). Unlike other social formations (like gender and race), they assert that Marxism "has been called upon to transform something that cannot be transformed" (ibid., p. 252). Regardless of geographical location, political system or cultural context, capitalism has been framed as a "post-ideological" force of nature that structures the possibilities for existence in our times (Schìld, 2019), "aims at the totalisation of its social relations" (Lacher, 2006, pp. 103–104), colonizes pre-capitalist and non-capitalist spaces (Hesketh, 2016) and has even converted "the distinctly political character, meaning, and operation of democracy 's constituent elements into economic ones" (Brown, 2015, p. 17). Contemporary social theory and cultural criticism teem with descriptions of the fantastical and destructive forces of capitalism as it consumes and commodifies all social relations and realities. Despite the radical intentions of such scholarship, the "assumption of subsumption" and overwhelming focus "on capital itself (how capitalism can be defined, how capitalism exploits, how capitalism survives and grows, how capital interacts and solidifies feudal modes of exploitation)" gives little space to imagine, let alone create, the possibilities for alternate economic relations (Hesketh, 2016, p. 881).

In order to subvert capitalism's perceived "largeness" and "proliferate possibility, not foreclose it" (2006b, p. 126), Gibson-Graham call for modes of inquiry and activism that disrupt capitalism's representation as "unified, singular and total" by recasting it as a dominant discourse (as opposed to a social structure) *and* engaging in the correlate project of creating the conditions for the excavation and creation of hitherto suppressed discourses of economic diversity (Gibson-Graham, 2006a, p. 259). They also contend that emancipatory projects must divorce ideas of economic transformation from projects of systemic transformation, and focus instead upon local, proximate and present alternatives to and variations among capitalism with the aim being "to create an approach to sites of socioeconomic activity that is dialectic rather than dualistic" (Hesketh, 2016, p. 879).

Neoliberalism as a crisis of imagination

> It seems to be easier for us today to imagine the thoroughgoing deterioration of the earth and of nature than the breakdown of late capitalism; perhaps that is due to some weakness in our imaginations.
>
> (Jameson, 1994, p. xii)

Another provocative line of critical economic inquiry favors viewing the crisis of neoliberal capitalism not as a crisis of policy or implementation but rather as a crisis of imagination. Max Haiven's *Crisis of imagination, crisis of power* asserts that capitalism[14] relies not only on the brutal repression of workers but also "on conscripting our imaginations" into conceiving of people as isolated, competitive economic agents; the natural world (including the human body) as an exploitable and renewable commodity; and capitalism as "the natural expression of human nature, or too powerful to be changed, or that no other system could ever be as desirable" (2014, p. 7). Describing the magnitude of the problem and the essential role of the imagination, Haiven writes, we "face an integration and exercise of capitalist power on a global scale that is both more widespread and more intensive than ever" (ibid., p. 30). Wielding the weapons of patriarchy, white supremacy, neocolonialism, ecological dislocation and dispossession, "capital increasingly reaches deep into our social, private, and subjective lives" (ibid., p. 30). *Crisis of imagination, crisis of power* explores the consequences of neoliberal capitalism's 50+ year recalibration of most spheres of life; most perversely, its transmutation of what and who we value and its correlate enclosure, privatization, and commodification of the communal sites and sources of imagination ("the commons") including public goods (e.g. water, food, shelter, transportation, childcare, healthcare, education, the environment), public spaces and institutions, as well as, the ideational commons of memory and creativity.

On the crisis of values Haiven asserts,

> systems of power work, in part, by conscripting our imaginations and misinforming our sense of value. What we imagine to be of value affects how we act and react to each other, to ourselves, to the world. Our actions, in turn, impact and inform our own imaginations and the imaginations of others.
>
> (ibid., p. 250)

As a social formation born *in ideological opposition to* the various socialist, democratic and decolonial movements emerging in the postwar period, early advocates of neoliberalism appropriated the political ideology of liberalism—and its values of freedom, equality and individualism—in the service of capital, conflating neoliberalism with democracy, and positioning it as the natural, inevitable and desirable outcome of historical progress. Since the 1970s we have seen the gradual subjugation of all social values under economic (monetary) value in a system that privileges production over reproduction, accumulation over distribution, growth over sustainability, competition over cooperation, and consumption over creativity. Haiven asserts that the more recent processes of "financialization" have effectively reimagined society in financial terms, producing and reproducing the power of "imaginary wealth" (and its very real profiteers) and "interpreting our economic, social and political lives as a portfolio of investments and speculative gambles" (2014, p. 106).

In order to reimagine value more deeply, Haiven asserts that we must both disentangle the question of social values from the question of economic value and learn to speak about values in compelling ways that are *not based in ideological oversimplification* or enmeshed in neoliberal ideologies and logics. Based as they are in the neoliberal values of growth and profit and in superficial commitments to ideological catchphrases, many contemporary solutions (e.g. the redistribution of wealth, green growth, sustainable development) will at best delay or peripheralize the crisis, and at worst, reinforce the very cycles of exploitation and destruction that they claim to redress.

According to Vandana Shiva, while the first enclosures of the commons in England "stole only land, today all aspects of life are being enclosed—knowledge, culture, water, biodiversity, and public services such as health and education" (2005, p. 3). Defining the commons both as "the real, existing alternative anti-capitalist institutions that make life worth living (community gardens, housing cooperative, social movements) and to the quality or timbre of many areas of our lives that we hold in common" (2014, p. 23), Haiven details neoliberalism's near total enclosure of all means of "imagining otherwise" (ibid., p. 25). The enclosure of memory is particularly important for the ways in which we remember the "radical events" of the past not as distinct and discrete events but rather, as "living products and producers of collective action" that thread the past, present and future (ibid., p. 25). Contemporary "histories" more often than not either write out, peripheralize, belittle, and defame radical events and communities of the past, or tokenize them as "the private property of heroic individuals" (ibid., p. 159). Such an approach to history both severs our connection to our (radical) forebearers and decommons activism. What is more, the enclosure of our common pasts has both commodified and defanged traditional forms of knowledge as quaint (if not backwards) relics of our premodern, prescientific past.

He also highlights the contemporary transformation of "creativity" from a collective act of making together into a big business notings that the commodification and de-commoning of creativity represents an extension of the Enlightenment transmutation of creativity from a "social, socialized, and socializing phenomenon" to a product and property of the "creative genius" (Haiven, 2014, p. 192). Not only has this seen the incredible growth of the so-called "cultural" and "creative industries," it has also seen "creativity" appropriated as an economic tool in most sectors. This new hype around creativity, however, obscures the fact that "we are ever more limited in our collective creative capacity to shape our destinies" (Haiven, 2014, p. 192). Moreover, the privatization and commodification of creativity actually "makes us less creative in how we think about social problems and solutions" (ibid., p. 192).

Marnie Holborow's (2015) *Language and neoliberalism* reveals the insidious ways in which the "market metaphor" has also commodified language itself asserting that the emergence of language skills as resources and sources of profit for the market parallels the rise of a global economy of information and communication and so-called knowledge work; what Dean (2014) has termed "communicative capitalism." Bringing this topic to bear on her own sector, Holborow dedicates a chapter to the "enthusiastic embrace" of market metaphors in Irish higher education focusing upon the contemporary flourish (and non-reflexive use) of the neoliberal keyword "entrepreneur" and its offshoots "enterprise" and "entrepreneurship" in university policies, missions and business plans. Building on Raymond William's core proposition in his study of ideological keywords (1985), Holborow reminds us that "the persuasive power of neoliberal keywords lies precisely in their apparent non-ideological character and their matter-of-fact, common-sense status as mere pointers to adaptation and advancement in market society" (2015, pp. 71–72).

The problem with work in capitalism

As stated in our introduction, labor is fraught in the capitalist imagination: while it is narrowly conceived in economics as referring only to wage labor, it constitutes *the* social scaffolding on which our characters are built, our incomes are determined and our daily lives are enacted (Livingston, 2016). As opposed to delivering on the promises of growth and progress, the past 50 years of neoliberal policy and reform have seen increased "precariousness, drudgery and anxiety" in work and life (Haiven, 2014, p. 300).

Following on the heels of the 2007/8 financial collapse, the COVID crisis has sharply illustrated (and exacerbated) several enduring but often latent themes about work in the contemporary economic system: First, we live in what various commentators have termed a "work society" (Weeks, 2011), described as a pervasive cultural preoccupation with (waged) work as an "ethic," form of (moral) character development and source of subjective identification (Livingston, 2016; Parrique, 2019). Recent perversions of this include workers becoming brand ambassadors for their organizations, being expected to donate financially in fundraising efforts (as in the case of our sector of higher education) and in the related rhetorical gesture by institutions that employees are "family"—a trend which has seen the recent meme "the institution does not love you back." Second, many people do not like their work or workplaces—from "shit jobs" (Graeber, 2018) to "bullshit jobs" (Graeber, 2013) to everything in between—many people are fed up with the forms of exploitation and alienation associated with their jobs and workplaces and are seeking alternatives to the capitalist social convention of life-as-wage-labor[15] that may involve going "off the grid" and/or exploring forms of self and cooperative employment (Alliger, 2021).

Part of the issue with work in a "work society" is its basis in an understanding of the employment relationship not as a social (and thus political) institution but rather as a unique relationship between employer and employed (Weeks, 2011). According to Weeks, work's historically vexed position "within the private–public division became even more troubled with the advent of industrialization; as work became identified with waged work and separated from the household" (ibid., p. 3). Marx's shift of the field of analysis from market exchange to factory exploitation functioned

> not only to publicize but also to politicize the world of work. That is to say, the focus on the consumption of labor seeks to expose the social role

> of work and, at the same time, to pose it as a political problem. Despite Marx's insistence that waged work for those without other options is a system of "forced labor" it remains for the most part an abstract mode of domination. In general, it is not the police or the threat of violence that force us to work, but rather a social system that ensures that working is the only way that most of us can meet our basic needs.
>
> (ibid., p. 7)

This violence is increasingly experienced in all forms of work and workplaces. According to Graeber, the vast majority of the workforce is now engaged in working long hours in jobs that they believe are not necessary about which he states: "This is a profound psychological violence here. How can one even begin to speak of dignity in labour when one secretly feels one's job should not exist? How can it not create a sense of deep rage and resentment" (2013, p. 5).

A final theme involves the growing recognition that

> capitalist society harbours a deep-seated social-reproductive "crisis tendency" or contradiction: on the one hand, social reproduction is a condition of possibility for sustained capital accumulation; on the other, capitalism's orientation to unlimited accumulation tends to destabilize the very processes of social reproduction on which it relies.
>
> (Fraser, 2015, p. 159)

O'Connor (1994) extends this contradiction asserting that capitalist accumulation and the market economy:

> depend for their viability on the exploitation not just of labour power in industrial production and of the physical habitat through raw material appropriation and waste disposal, but also on exploitation of noncapitalist social domains and publicly provided infrastructures. These non-commodity domains, the domains of intertwined human, communal and natural activities of repair, renewal, regeneration, and reproduction, may thus be thought of as furnishing the necessary material and social conditions for commodity production, market exchange, and capital accumulation. Moreover, capital, whether offshore or with particular territorial affiliations, provokes, through its own parasitism, a "crisis" in the reproduction of these needed conditions of production.
>
> (p. 108)

Haiven reminds us that even the so-called "heyday" of capitalism in the 1950s–1960s was built on a white, patriarchal middle class normativity that depended upon the subjugation of women, minorities and the working class and the correlate neocolonial exploitation of the Global South. Barca (2020) situates these subjugations in the four dualisms that form the basis of capitalist relations of power: the male–female dualism that forms the basis of gendered and sexed divisions of labor and the expropriation of reproductive labor; the mind–body dualism that forms the basis of the classed, property relations and the division between intellectual and manual labor; the human–nature dualism that forms the basis for the expropriation of nature in economic activities; and the civilized–savage[16] dualism that forms the basis of colonial and raced relations and the expropriation and exploitation of marginalized people's labor (see also Plumwood, 1986). Barca (2020) states that all four dualisms are brought together in the production–reproduction dualism, which she asserts is at the heart of our ecological crisis.

While Marx did not specifically address reproductive labor, he did argue that the exclusive focus on market exchange in bourgeois political economy did not adequately account for the "social metabolism" involved in capitalist modes of production (Murillo and D'Atri, 2018). Feminist Marxists on the other hand have long studied forms of work that are systematically devalued in capitalist systems, especially those forms that are denied the denomination of work altogether as in the case of reproductive and/or care labor (Bhattacharyya, 2018). Marxist feminism defines social reproduction as "the activities necessary to maintain and reproduce life daily and intergenerationally at both the individual and social scale" (Winders and Smith, 2019, p. 872). Winders and Smith's (2019) historical survey of the field identifies four feminist imaginaries of social reproduction (theory) as they articulate the crisis of reproduction at the center of the capitalist economy. The first imaginary, associated with the work of Mariarosa Dalla Costa and Selma James in the 1970s, problematized the "separate and (un?)equal status of reproductive labor" in the capitalist economy and saw the formation of (with Silvia Federici and Brigitte Galtier) the International Feminist Collective which both publicized this contradiction and coordinated action networks around the idea of "wages for housework" (see Dalla Costa, 1972; Dalla Costa and James, 1975). In the context of women increasingly entering the workforce in 1980s (related to declining real wages and deindustrialization), we see a shift in the feminist imagination to productive–reproductive labor as an "overlapping duality of equals" (Winders and Smith, 2019, p. 876). This imaginary was influenced greatly by the work of feminist geographers (e.g. Pratt and Hanson, 1993; Katz, 2008; England, 2010; Peck, 1996) who, noting the splintering of the neatly bounded separation of work and home into a proliferation of household arrangements and strategies, "envisioned dynamic interdependencies between work and home" foregrounding women's socially necessary activities in empirical terms (Winders and Smith, 2019, p. 877).

The third feminist imaginary of social reproduction problematized the contradiction between the "degendered and unraced figure" living at the center of the neoliberal economy and its ironic dependence on "deeply gendered (and classed and racialized) dynamics" (Winders and Smith, 2019, p. 879; see also England, 2010). Scholarship in this period notes the varied ways in which the distinction between the space to produce and reproduce has become completely blurred. They assert, "not only did work and life converge spatially they occurred simultaneously" (p. 879). Reflecting the variegated nature of neoliberal capitalism in this period, scholars began drawing attention to the need for research that explores the different ways social reproduction has been threatened and exploited in different parts of the world (Kunz, 2010; Shiva, 2005; Salleh, 2017). As Katz (2001) contends, the conditions for social reproduction are always in dialectical relation with production and as such are consistently restructured as capitalist systems shift to new geographies to exploit new relations of production and accumulation. The fourth imaginary of social reproduction explores the ways in which social reproduction is increasingly distributed via an international division of labor, where migrant women from the Global South provide reproductive and care labor for families in the Global North. This imaginary also provides a conceptual framework for reflecting upon labor that, while paid and productive, has reproductive components or functions, what Nancy Folbre (2014) and Nancy Fraser (2016) refer to as caring and/or care labor.

While social reproduction provides a useful framework for theorizing and empirically studying reproductive labor, most analyses still focus on the dualisms of home–work and male–female as the grounds for struggle, when it is clear that the global crises of production and reproduction we now face necessitate an intersectional approach to reproduction that accounts for the variegated and uneven ways in which reproductive sources ("commons") are exploited, expropriated and resisted. Rodríguez-Rocha asserts that the inclusion of poststructuralist frameworks including intersectionality theory and post/de-colonial knowledges from the global periphery can provide much needed insights into the "material, subjective and lived experiences of multiple oppressions and organized resistance under capitalism as they bear on the practices and organization of social reproduction" (2021, p. 11). Di Chiro (2008) adds the need for approaches that address environmental problems as problems of capitalist social reproduction. According to Haiven and Khasnabish, the neoliberal crisis of social reproduction has become a wholesale global crisis of social reproduction, where social life is made to pay the cost of the reproduction of a renegade capitalist system even as those lives are increasingly unable to reproduce themselves" (2014, p. 11).

The following chapter introduces our theoretical approach to the study of the crisis of neoliberalism—social imaginaries—as it provides a useful (and hopeful) framework for reimagining economic relationships and institutions and crucially reimagining labor on more ethical, sustainable and convivial grounds.

Notes

1 Ayn Rand's ideas regarding the value of selfishness continue to influence neoliberal thinking.
2 F.A. Hayek's coinage.

3 English use of the previously existing French term, "néo-libéralisme," is credited first to French economist Charles Gide who used it in 1898 to describe the economic beliefs of the Italian economist Maffeo Pantaleoni and second to the 1938 "Colloque Walter Lippmann" a conference of intellectuals organized in Paris where the term neoliberalism was chosen as the preferred term to describe (and advocate for) a certain set of economic beliefs. There was ultimately division within the group over the role of the state, and the Colloque ultimately languished as attentions turned to the First World War. Crucially, Fredrich Hayek was a central member and he brought these ideas to the Mont Pelerin Society. The term is variously associated with the intellectuals associated with the Austrian School of Economics (under Ludwig von Mises and F. A. Hayek) and the German Freiburg School (where Hayek also eventually taught).

4 Though signed by 56 nations, the ILO's founding document, the Havana Charter, (proposed by economist John Maynard Keynes) ultimately failed because the US Congress refused to submit it for ratification on the grounds that it would intervene in internal economic affairs. Many believe the treaty's commitment to "full employment" was pivotal in explaining its rejection.

5 There were no female founders to our knowledge.

6 Chang (2001) calls neoliberalism the "unholy alliance" between neoclassical economics and the Austrian Libertarian tradition as he points to Hayek's 1949 essay criticizing the neoclassical school of thought.

7 According to the society's "About" page, George Stigler once quipped that the Society could be called "The Friends of F. A. Hayek" (www.montpelerin.org/about-mps/)

8 It is worth mentioning that Hayek was also closely associated with the Chicago School.

9 Labor successes saw increased wages, increased safety standards for the working class and higher production costs for the capitalist class.

10 There is no denying that the manipulation also went the other way and these pro-social movements were used by authoritarian socialist state leaders to destabilize many Western European and North American political regimes.

11 John B. Thompson (2001) asserts that theories of "new social movements" of that period underestimate and misrepresent the actual work and consequences of these movements in prefiguring "other worlds".

12 For example, US President Reagan (who ironically supported the Screen Actors Guild when he was a working actor) famously fired striking Air Traffic Controllers in 1981 and UK Prime Minister Thatcher dissolved the worker's union of miners following their 1984 strike.

13 Even so "many at the IMF, World Bank and in commanding positions continue to trumpet the supposed benefits of trade liberalization, open borders, and financial integration, more recently in the face of resurgent currents of protectionism, nativism, and nationalism in the United States and in parts of Europe" (Peck et al., 2018, p. xxiii).

14 Haiven uses both "capitalism" and "neoliberalism."

15 The contemporary resurgence of "maker culture" (e.g. craft and DIY communities and hacker and makerspaces) is a noteworthy manifestation of resistance to the neoliberal commodification of creativity and its correlate emphasis on the act of consumption. In their various forms, maker practices and communities are a part

of a larger cultural opposition to the commodification, exploitation and enclosure of the act of creating (see Gauntlett, 2011; Vogelaar and McKernan, 2018).

16 For Marx and Engels the primary, decisive division is that between town and country. According to Magdoff (2006) the town-country antithesis encompasses much more than city vs. farm. Today, even in the most advanced industrial countries, conflicts and contrasts exist between, on the one hand, regions that specialize in industry, commerce and finance, and, on the other, those that engage primarily in agriculture. Furthermore, with the progress of international trade and empire-building by the industrially and militarily superior capitalist nations, an international division of labor is created and reproduced (by the use of force and the "normal" operations of the market) between the core countries ("town") and those of the periphery ("country") (https://monthlyreview.org/2006/10/01/the-meaning-of-work-a-marxist-perspective/).

References

Alliger, G.M. (2021) *Anti-work: Psychological investigations into its truths, problems, and solutions*. New York: Routledge.

Barca, S. (2020) *Forces of reproduction: Notes for a counter-hegemonic Anthropocene*. Cambridge: Cambridge University Press.

Best, J. (2020) The quiet failures of early neoliberalism: from rational expectations to Keynesianism in reverse. *Review of International Studies*, 46(5), pp. 594–612.

Bhattacharyya, G. (2018) *Rethinking racial capitalism: Questions of reproduction and survival*. Lanham: Rowman & Littlefield.

Birch, K. and Springer, S. (2019) Peak neoliberalism? Revisiting and rethinking the concept of neoliberalism. *Ephemera Journal*, 19(3), pp. 467–485.

Boas, T.C. and Gans-Morse, J. (2009) Neoliberalism: From new liberal philosophy to anti-liberal slogan. *Studies in Comparative International Development*, 44(2), pp. 137–161.

Brenner, N., Peck, J. and Theodore, N. (2010) Variegated neoliberalization: Geographies, modalities, pathways. *Global Networks*, 10(2), pp. 182–222.

Brown, W. (2015) *Undoing the demos: Neoliberalism's stealth revolution*. Cambridge: MIT Press.

Cahill, D., Cooper, M., Martijn Konings, M. and Primrose, D. (eds.) (2018) *The Sage handbook of neoliberalism*. London: Sage.

Castree, N. (2010) Neoliberalism and the biophysical environment 1: What "neoliberalism" is, and what difference nature makes to it. *Geography Compass*, 4(12), pp. 1725–1733.

Chang, H. (2001). Breaking the mould: An institutionalist political economy alternative to the neoliberal theory of the market and the state. Social Policy and Development Paper Number 6, United Nations Research Institute for Social Development.

Cornelissen, L.S. (2019) On the (ab) use of the term "neoliberalism": Reflections on Dutch political discourse. Ephemera: *Theory & Politics in Organization*, 19(3), pp. 487–512.

Dalla Costa, M. (1972) *Potere femminile e sovversione sociale*. Venice: Marsilio Editori.

Dalla Costa, M. and James S. (1975) *The power of women and the subversion of the community*. London: Falling Wall Press.

Dean, J. (2014) Communicative capitalism and class struggle. *Spheres: Journal for digital cultures*, 1, pp. 1–16.

Di Chiro, G. (2008) Living environmentalisms: Coalition politics, social reproduction, and environmental justice, *Environmental Politics*, 17(2), pp. 276–298. doi:10.1080/09644010801936230

England, P. (2010) The gender revolution: Uneven and stalled. *Gender & Society*, 24(2), pp. 149–166.

Fisher, M. (2009) *Capitalist realism, is there no alternative?*. Winchester: O Books.

Flanders, L. (2013) At Thatcher's funeral, bury TINA, too. *The Nation*, April 12. Available at: www.thenation.com/article/archive/thatchers-funeral-bury-tina-too/

Folbre, N. (2014) The care economy in Africa: Subsistence production and unpaid care. *Journal of African Economies*, 23(suppl_1), pp. 128–156.

Fraser, N. (2015) Legitimation crisis? On the political contradictions of financialized capitalism. *Critical Historical Studies*, 2(2), pp. 157–189.

Fraser, N. (2016) Capitalism's crisis of care. *Dissent*, 63(4), pp. 30–37.

Galbraith, J.K. (2021) The death of neoliberalism is greatly exaggerated. *Foreign Policy*, April 6. Available at: https://foreignpolicy.com/2021/04/06/death-neoliberalism-larry-summers-biden-pandemic/

Gauntlett, P. (2011) *Making is connecting: The social meaning of creativity from DIY and knitting to YouTube and Web 2.0*. Cambridge: Polity Press.

Garretón, M.A. (2004) *Incomplete democracy: Political democratization in Chile and Latin America*. Durham: UNC Press Books.

George, S. (2004) *Another world is possible if …* London: Verso.

Gibson-Graham, J.K. (2006a) *The end of capitalism (as we knew it): A feminist critique of political economy*. Minneapolis: University of Minnesota Press.

Gibson-Graham, J.K. (2006b) *A postcapitalist politics*. Minneapolis: University of Minnesota Press.

Graeber, D. (2008) Hope in common. *The anarchist library*. Available at: https://theanarchistlibrary.org/library/david-graeber-hope-in-common

Graeber, D. (2013) On the phenomenon of bullshit jobs: A work rant. *Strike Magazine*, 3, 1–5.

Graeber, D. (2018) *Bullshit jobs*. New York: Simon and Schuster .

Harvey, D. (2005). *A brief history of neoliberalism*. Oxford: Oxford University.

Haiven, M. (2014) *Crises of imagination, crises of power: Capitalism, creativity and the commons*. London: Zed Books.

Haiven, M. and Khasnabish, D.A. (2014) *The radical imagination: Social movement research in the age of austerity*. London: Bloomsbury Publishing.

Hesketh, C. (2016) The survival of non-capitalism. Environment and Planning *D: Society and Space*, 34(5), pp. 877–894. doi:10.1177/0263775816639313

Holborow, M. (2015) *Language and neoliberalism*. New York: Routledge.

Jameson, F. (1994) *The seeds of time*. New York: Columbia University Press.

Jessop, B. (2013) Putting neoliberalism in its time and place: A response to the debate. *Social Anthropology*, 21(1), pp. 65–74. doi:10.1111/1469-8676.120032

Jones, G. (2013) Afterword: Rates of exchange: Neoliberalism and the value of higher education. *International Studies in Sociology of Education*, 23(3), pp. 273–280.

Katz, C. (2001) Vagabond capitalism and the necessity of social reproduction. *Antipode*, 33(4), pp. 709–728.

Katz, C. (2008) Bad elements: Katrina and the scoured landscape of social reproduction. *Gender, Place and Culture*, 15(1), pp. 15–29.

Klein, A. (2007) Judging as nudging: New governance approaches for the enforcement of constitutional social and economic rights. *Columbia Human Rights Law Review*, 39, pp. 351–422.

Klein, M. and Pettis, M. (2021) *Trade wars are class wars*. New Haven: Yale University Press.

Kotz, D.M. (2015) Neoliberalism, globalization, financialization: Understanding post-1980 capitalism. *The restructuring of capitalism in our time*, Department of Economics, University of Massachusetts Amherst School of Economics. Available at: www.umass.edu/economics/sites/default/files/Kotz.pdf

Kunz, R. (2010) The crisis of social reproduction in rural Mexico: Challenging the "re-privatization of social reproduction" thesis. *Review of International Political Economy*, 17(5), pp. 913–945.

Lacher, H. (2006) *Beyond globalization: Capitalism, territoriality and the international relations of modernity*. New York: Routledge.

Livingston, J. (2016) Fuck work. Aeon. Retrieved on June 2, 2022 from https://aeon.co/essays/what-if-jobs-are-not-the-solution-but-the-problem

Loewenstein, A. (2017). *Disaster capitalism: How to make a killing our of catastrophe*. London: Verso.

Monbiot, G. (2016) Neoliberalism: The ideology at the root of all our problems. *The Guardian*, April 15.

Munck, R. (2003) Neoliberalism, necessitarianism and alternatives in Latin America: there is no alternative (TINA)?. *Third World Quarterly*, 24(3), pp. 495–511.

Murillo, C. and D'Atri, A. (2018) Producing and reproducing: Capitalism's dual oppression of women. *Left Voice Magazine*. Retrieved in May 2022 from www.leftvoice.org/on-reproductive-labor-wage-slavery-and-the-new-working-class/

O'Connor, M. (1994) The second contradiction of capitalism: The material/communal conditions of life. *Capitalism, Nature, Socialism*, 5(4), pp. 105–114. doi: 10.1080/10455759409358613

Parrique, T. (2019) *The political economy of degrowth* (Doctoral dissertation, Université Clermont Auvergne (2017–2020)).

Patnaik, P. (2007) The state under neo-liberalism. *Social Scientist*, 35(1/2), pp. 4–15.

Patnaik P. and Patnaik, P. (2016) *A theory of imperialism*. New York: Columbia University Press.

Peck, J. (1996) *Work-place: The social regulation of labor markets*. New York: Guilford Press.

Peck, J. (2013) Explaining (with) neoliberalism. *Territory, Politics, Governance*, 1(2), pp. 132–157.

Peck, J. (2018). Preface: Naming Neoliberalism. In Cahill, D., Cooper, M., Konings, M. and Primrose, D. (eds.). The SAGE handbook of neoliberalism. Los Angeles: SAGE Publications, pp. xxii–xxiv.

Peck, J., Brenner, N. and Theodore, N. (2018) Actually existing capitalism. In Cahill, D., Cooper, M, Konings, M. and Primrose, D. (eds.). *The SAGE handbook of neoliberalism*. Los Angeles: SAGE Publications, pp. xxii–xxxii

Phelan, S. (2007) Messy grand narrative or analytical blind spot? When speaking of neoliberalism. *Comparative European Politics*, 5(3), pp. 328–338.

Phelan, S. (2014) *Neoliberalism, media and the political*. Berlin: Springer.

Plumwood, V. (1986) Ecofeminism: An overview and discussion of positions and arguments. *Australasian Journal of Philosophy*, 64(sup1), pp. 120–138.

Pratt, G. and Hanson, S. (1993) Women and work across the life course. In Katz, C. and Monk, J. (eds.). *Full circles: Geographies of women over the life course.* London and New York: Routledge, pp. 27–54.

Rodríguez, J.P. (2021) The politics of neoliberalism in Latin America: Dynamics of resilience and contestation. *Sociology Compass*, 15(3). https://doi.org/10.1111/soc4.12854

Rodríguez-Rocha, V. (2021) Social reproduction theory: State of the field and new directions in geography. *Geography Compass*, 15(8). https://doi.org/10.1111/gec3.12586

Salleh, A. (2017) *Ecofeminism as Politics: Nature, Marx and the postmodern.* London: Zed Books Ltd.

Schild, V. (2019) Feminisms, the environment and capitalism: On the necessary ecological dimension of a critical Latin American feminism. *Journal of International Women's Studies*, 20(6), pp. 23–43.

Shiva, V. (2005) *Earth democracy-justice, sustainability and peace.* Berkeley: North Atlantic Books.

Springer, S. (2014) Neoliberalism in denial. *Dialogues in Human Geography*, 4(2), pp. 154–160.

Springer, S. (2016a) Fuck neoliberalism. *ACME: An International Journal for Critical Geographies*, 15(2), pp. 285–292.

Springer, S. (2016b). *The discourse of neoliberalism: An anatomy of a powerful idea.* Rowman & Littlefield.

Stedman-Jones, D. (2013) The American Roots of Neoliberalism. *History News Network.*

Stokes, S.C. (2001) *Mandates and democracy: Neoliberalism by surprise in Latin America.* Cambridge: Cambridge University Press.

The Mont Pellerin Society (1947) Statement of aims, April 8. Available at: www.montpelerin.org/statement-of-aims/

Toniolo, G. (1998) Europe's golden age, 1950–1973: Speculations from a long-run perspective. *Economic History Review*, 51(2), pp. 252–267.

Venugopal, R. (2015) Neoliberalism as concept. *Economy and Society*, 44(2), pp. 165–187.

Vogelaar, A. and McKernan, C. (2018). Making space for a revolution: Occupy Wall Street as a maker movement, in Hunsinger, J. and Schrock, A. (eds.). *Making our world: The hacker and maker movements in context.* New York: Digital Formations, Peter Lang.

Wacquant, L. (2012) Three steps to a historical anthropology of actually existing neoliberalism. *Social Anthropology/Anthropologie Sociale*, 20(1), pp. 66–79.

Weeks, K. (2011) The problem with work. In *The problem with work.* Durham: Duke University Press.

West, P. and Brockington, D. (2012) Introduction: Capitalism and the environment. *Environment and Society: Advances in Research*, 3(1–3). doi:10.3167/ares.2012.030101

William, R. (1985) Walking backwards into the future. *New Socialist*, 27, pp. 21–23.

Wilson, J.A. (2017) *Neoliberalism.* New York: Routledge.

Winders, J. and Smith, B.E. (2019) Social reproduction and capitalist production: A genealogy of dominant imaginaries. *Progress in Human Geography*, 43(5), pp. 871–889. doi: 10.1177/0309132518791730

3 Social imaginaries

> Every social process of production is at the same time a process of reproduction.
> (Marx, *Capital*, p. 71)

This book approaches the crisis of capitalism using the conceptual framework of "social imaginaries" as first formulated by philosopher and economist Cornelius Castoriadis in *The Imaginary Institution of Society* (1975). The area has been subsequently elaborated by scholars of global modernity including Benedict Anderson, Arjun Appadurai, Dilip Parameshwar Gaonkar, Charles Taylor, and members of the Center for Transcultural Studies (CTS), Manfred B. Steger and Paul James, and more recently, by a diverse set of scholars of social theory and change (e.g. Adams et al., 2015; Arnason, 2015; Asara, 2020, Haiven, 2014; Haiven and Khasnabish, 2014; Varvarousis, 2019). The emergence of the idea of social imaginaries is rooted in foundational debates in Western intellectual history (beginning with the Ancient Greeks) about the nature of reality and being, the relationship between reality and imagination, and the degrees to which human societies and identities are predetermined or autonomously created.

The imagination was given a new centrality in the context of the European Enlightenment of the eighteenth and nineteenth centuries when philosophers including David Hume, René Descartes and, at the extreme end, Immanuel Kant, theorized the (white, male, bourgeoise, individual) imagination as the (divinely bestowed) distinction setting (some) humans apart from animals. According to Haiven, "the Enlightenment idea of the imagination was to become a fetish by which European colonial and imperial regimes justified in part their domination of other civilizations" and other (non-white, non-male, non-bourgeoise, non-human) beings who were seen as both "too imaginative" and "not imaginative enough" (2014, p. 223). The imagination was central to that period's social upheaval; where on the one hand it was actively mobilized in the justification of European feudalism and colonialism; it was also a central tool in resistance and calls for radical social change as for example in the Romantic movement's notions of the imagination as central to resistance and revolution and in the emergent field of psychoanalysis. Karl

DOI: 10.4324/9781003138617-3

Marx's and Friedrich Engel's "historical materialism" was in part a reaction to Enlightenment overemphasis on the power of the individualistic, creative imagination and Max Weber's contribution to the conversation—that society and history are "overdetermined" (that is, the culmination of complex and undetermined forces that play out differently in different cultural contexts)—forms the basis of contemporary social imaginary research.

The study of "social imaginaries"[1] as such emerged in the context of the socio-political and economic transformations associated with the 1950–1960 postwar reconstruction period, surfaced again in the context of the "cataclysmic events" of the 1980s–1990s (and the often violent victories of "liberal democracy" (and capitalism) over socialism in the post-Cold War period) and more recently following the state-led resuscitation of neoliberal institutions after the 2008 global financial crisis (Gaonkar, 2002, p. 1). Indeed, the concept of social imaginaries has developed alongside and, in relation to, the development of capitalism and is directly related to Marxist thought.

The concept of imaginaries is connected to broader—critical, discursive, hermeneutic, poststructural and psychoanalytic—turns in the humanities and social sciences in the latter half of the twentieth century. These "turns" variously challenged the dominant materialist, essentialist and positivist approaches to the study of social institutions and processes and broadly investigated the dynamic psychological and social processes through which individuals and societies are constituted, controlled and changed. Castoriadis's notion of the "social imaginary" emerged in opposition to materialist and determinist conceptions of society and history in twentieth-century Marxist thought.[2] Against this grain, Castoriadis asserted that each society is organized around (different) stable but shifting social imaginaries that emerged from specific social, spatial and historical contexts, manifested in institutions, subjectivities, practices and relations and were constituted (and crucially *made viral*) via the symbolic realm. While Castoriadis agreed with Marx that the history of capitalism is ridden with crisis and contradiction, he did not believe that future historical developments (e.g. revolution) could be deduced from them. Instead, he believed that the outcome of capitalist crises would be "determined by how individuals and society *take up* those crises, not by any necessary, internal self-development of capitalism 'itself'" (Garner, nd).

Castoriadis counteracted the reigning "ontology of determinacy" by theorizing the creative force underlying all social formations and theorizing the ideal conditions under which it was cultivated. He distinguished between two ideal types of socio-historical formations: heteronomous and autonomous. Whereas in heteronomous societies, laws, norms, values, myths and meanings were understood as given once and for all, preordained from some eternal, external source, in autonomous societies these social formations are understood as contingent and constructed (Gaonkar, 2002). According to Castoriadis, the reflexivity and creativity central to an autonomous society were "made possible through the simultaneous invention and institution of [Western style] philosophy and democracy" (Gaonkar, 2002, p. 9).

For Castoriadis, the social imaginary is a generative force; the shared collective imagination embodied in specific institutions (tools, language, skills, norms, values, etc.) which operates as the "invisible cement" holding a society together and determining "what is real, worthy, possible, acceptable or desirable" (Varvarousis, 2019, p. 7). The existence of a dominant social imaginary does not preclude the existence of other social imaginaries; in any given society even groups with different, peripheral imaginaries "can recognize the existence of central social imaginary significations and grasp the meaning of them—but not necessarily adopt them" (Varvarousis, 2019, p. 7). Indeed, it is this very plurality and contradiction that opens up the terrain to contestation and play, to the idea that another imaginary is possible.

In spite of Castoriadis's enduring and profound contribution to economic and social inquiry, Gaonkar rightly notes its limitations, including its "staggering Eurocentrism [...] and its idealization of ancient Greece," its highly abstract level of ontological reflection that "rarely engages the question of how change and difference are produced locally through the workings of the social imaginary's significations at specific social historical conjunctures" as well as its subsumption of societies into a heteronomous-autonomous dichotomy that does not accommodate the fact that "all social formations, at least the modern ones, differentially incorporate aspects of both" (2002, p. 9). Subsequent elaborations of social imaginaries have addressed one or more of these important limitations, extending Castoriadis's ideas into the realm of global modernity and its correlate problems and possibilities.

Modern social imaginaries

The concept and study of social imaginaries is also associated with a preoccupation in social theory with what Charles Taylor (2004) has asserted is the number one problem facing modern social science, that is:

> modernity itself: that historically unprecedented amalgam of new practices and institutional forms (science, technology, industrial production, urbanization), of new ways of living (individualism, secularization, instrumental rationality), and of new forms of malaise (alienation, meaninglessness, a sense of impending social dissolution).
>
> (p. 1)

In addition to providing a framework for interrogating the novel forms of being and belonging associated with modernity, the language of "imaginaries" gestures at a central concern with the universalizing and colonizing tendencies of modern social formations and the novel ways in which they exert "power through" as opposed to "power over." As such, social imaginary research assumes a poststructural (even anti-structural) posture towards culture and society as deeper than rules and structures and as constituted by webs of meaning that individuals do not merely decode and follow but rather embody and enact.

The contributions summarized below have set the stage for a rich, complex and diverse "paradigm-in-the-making" that is inherently eclectic and interdisciplinary and predominantly theoretical (Adams et al., 2015). The term has in recent years become a "buzzword" surfacing regularly in academic essays addressing a wide range of (global) social, economic, political and environmental issues. The field of inquiry is indebted to a long list of scholars working in many disciplines across the world; a list that is simply too long to adequately address in this chapter. We offer here some significant contributions to the field, each of which opens a door to a others.

Contemporary conceptualizations of the social imaginary are indebted to sociologist Benedict Anderson's (1983) formulation of the nation as an imagined community in his seminal *Imagined Communities*. Anderson's interpretation emphasizes the central role of communication in the formation of imaginaries asserting that "national consciousness" was an effect of the interplay between technology (print media), capitalism (as a system of production and productive relations) and the related death of linguistic diversity. He asserts that the convergence of capitalism and print technology (print-capitalism) created the need for print languages in order to create economies of scale, the use and spread of which created the three key conditions for nationalism: unified fields of communicative exchange, language fixity and a new and more accessible "language of power." In addition to highlighting the central role of communication (construed as both language/symbolic representations and technologies) in the formation and mediation of social imaginaries, Anderson's "three paradoxes of nationalism" adeptly capture the paradox of all modern imaginaries (and indeed the social theories that attempt to account for them), namely: though they emerge in specific historical contexts they are imagined as eternal, they oscillate in the modern tension between the universal and the particular, and they exert concrete power even as they are hard to pin down.

Anthropologist Arjun Appadurai's (1990) "Disjuncture and difference in the global cultural economy," takes on these paradoxes in the context of the rapidly globalizing, hyper-capitalistic period of the 1980s–1990s. Responding to the intellectual dangers of "obliterating difference within the 'third world', eliding the social referent [common in postmodern theory] and retaining the narrative authority of Marxist tradition" (p. 308), Appadurai asserts that "the new global cultural economy has to be understood as a complex, overlapping, disjunctive order, which cannot any longer be understood in terms of existing [models]" (ibid., p. 296). He offers the framework of global scapes—ethno-, techno-, finance-, media, and ideo—as the landscapes through which diverse "imagined worlds" are constructed. The language of "scapes" highlights the fluid and positional nature of such constructs as they are inflected by the "historical, linguistic and political situatedness of different sorts of actors" from the nation-state to the individual (and the variety of sub and supra national groupings in between). In addition to decentering modern social imaginaries and their study, Appadurai's "global flows" position them as subjective sites of play and contestation mobilized by diverse interests and groups.

Modernity at Large (1996) deepens and expands Appadurai's exploration of the joint effect of media and migration "on the work of the imagination as a constitutive feature of modern subjectivity" (p. 2). The concept of "deterritorialization" features centrally has it helps describe the "disjunctures" and "conjunctures" that characterize global modern identities and communities. His own experiences in the 1960s

> drifting from one sort of postcolonial subjectivity (Anglophone diction, fantasies of debates in the Oxford Union, borrowed peeks at Encounter, a patrician interest in the humanities) to another: the harsher, sexier, more addictive New World of Humphrey Bogart reruns, Harold Robbins, Time, and social science, American-style

sets the foundation and provides conceptual fodder for the book's case studies. Its concluding chapter, "The production of locality" remains as salient today as it was in 1996. The chapter begins by reframing the ethnographic project, about which he states:

> drawn into the very localization they seek to document, most ethnographic descriptions have taken locality as ground not figure, recognizing neither its fragility nor its ethos as a property of social life. This produces an unproblematized collaboration with the sense of inertia on which locality, as a structure of feeling, centrally relies.
>
> (ibid., p. 182)

He asserts to the contrary that locality is, and has always been, "an inherently fragile social achievement [...] that must be maintained carefully against various kinds of odds" (ibid., p. 179). His exploration of the production of locality in delocalized societies recognizes the "the neighborhood" as the actually existing, and inherently colonizing, social form in which locality is variably realized. The chapter leaves us with a series of questions about the consequences of the increased efforts of the nation-state to "define all neighborhoods under the sign of its forms of allegiance and affiliation," the growing disjunctures associated with deterritorialization and the "steady erosion, principally due to the force and form of electronic mediation, of the relationship between spatial and virtual neighborhoods" (ibid., p. 189).

Alternative modernities

Rhetorician Dilip Parameshwar Gaonkar's edited volume, *Alternative Modernities* (2001), brings together key thinkers on the emergent line of inquiry exploring modernity as a global social formation that is fundamentally plural and fragmented. As in *Modernity at Large*, the studies in this book are guided by an interest in what difference site-based readings of modernities make in our understanding and questioning of the present. According to Gaonkar, "to think in terms of alternative modernities is to privilege a particular angle of interrogation"—in this case national/cultural sites—as it complicates our understanding of the tale of two modernities (societal modernization and cultural modernity) and demonstrates that tale's inability "without important modifications, to cover other theaters of modernity" (2001, p. 14). According to the author, thinking through and against Western modernity's self-understandings is to think with "a difference that would destabilize the universalist idioms, historicize the contexts, and pluralize the experiences of modernity" (ibid., p. 14). Contributions to this volume explore site-specific "creative adaptations" that, taken together, disrupt the prevailing "acultural" theory of modernity (as articulated by Charles Taylor, 1999) in favor of a cultural theory viewing modernity "not as one, but many."

Gaonkar's (2002) introduction to the special issue of *Public Culture*, "Towards new imaginaries," describes the conceptual turn toward social imaginaries among members of the Center for Transcultural Studies (CTS), a central group in the "alternative" and "multiple modernities" movement. In 1999, members drafted a statement on "new imaginaries" consisting of five core ideas:

> First, social imaginaries are ways of understanding the social that become social entities themselves—mediating collective life. [...]. Second, modernity in its multiple forms seems to rely on a special form of social imaginary that is based on relations among strangers [...] Third, the national people is a paradigmatic case of modern social imaginary. Its distinctive features include its representation as a "we"; its transparency between individual and collectivity; its agential subjectivity, in which a people acts in time; its unfolding in progressive history; and its posited environment of mutuality with other national peoples [...] Fourth, a national people lives amid many other social imaginaries, penumbral to them [...] Fifth, the agency of modern social imaginaries comes into being in a number of secular temporalities rather than existing eternally in cosmos or higher time.
>
> (pp. 4–5)

Fleshing out the key insights (and limitations) of Cornelius Castoriadis's work, this introduction (and the chapters that follow) highlight the "radically different intellectual and political milieu" underlying this new generation of social imaginary research. Core among the differences is its relationship with "the cataclysmic events of 1989 and their aftermath" including: the demise of the Soviet Union and "liberation" of Eastern Europe, rise of democratic movements in Asia and Latin America (and the correlate defensiveness of authoritative regimes across the world), emergence of new social movements, and revived interest in civil society and the public sphere (with international inflections) (ibid., p. 1). Picking up where Castoriadis left off, the contributions in this special issue take on "multiplicity" not as the answer to the philosophical riddle (of determinacy) but, "as a riddle itself" (ibid., p. 10). Gaonkar highlights CTS member Charles Taylor's insights on social imaginaries (detailed below), notably his suggestion that imaginaries function not only as a generative matrix, but also "in the hermeneutics of everyday life" (ibid., p. 10).

Charles Taylor's (2004) magnum opus, *Modern Social Imaginaries,* fleshes out his continuing engagement with the notion of "multiple modernities," defined as

> the plural reflecting the fact that other non-Western cultures have modernized in their own way and cannot properly be understood if we try to grasp them in a general theory that was designed originally with the Western case in mind.
>
> (p. 1)

Taylor's definition of social imaginaries (a term he adapted from Bronislaw Baczko's 1984 *Les imaginaries sociaux*) as "the ways people imagine their social existence, how they fit together with others, how things go on between them and their fellows, the expectations that are normally met, and the deeper normative notions and images that underlie these expectations" (ibid., p. 23) emphasizes the centrality of "self-understandings". Taylor's extensive exploration traces the origin and *mutation* of a new (liberal) moral order of society (based upon natural rights, freedom, mutual benefit and an elastic notion of equality) beginning in seventeenth-century Europe on its "long march" from theory to social imaginary. Central to this march was the "great disembedding"—a process of secularization that gave rise to new ideas of self and society that formed the basis for modern social formations including the market economy, public sphere and democratic governance. Addressing what he sees as a false dichotomy between "ideas and material factors as rival causal agencies," Taylor asserts that "what we see in human history is ranges of practices that are both at once, that is, material practices, carried out by human beings in space and time, and very often coercively maintained, and at the same time, self-conceptions, modes of understanding" (ibid., p. 31).

The Eurocentric bias in social imaginary research is addressed in detail in Gurminder K. Bhambra's (2007), *Rethinking Modernity*, which offers a postcolonial critique centered on the politics of knowledge production and the Eurocentric theories of modernity at the heart of the modern "sociological imagination." Emerging as it did in the context of European colonialism and empire, modern social theory is besieged with critical omissions and misunderstandings. The most foundational of which are the imagined temporal rupture with an agrarian past and imagined distinctiveness of Europe at the heart of European notions of modernity. In spite of their attempts to decenter Western formations, Bhambra asserts that critical and postcolonial scholarship often remain complicit in the privileging of Europe as the ultimate reference point, "albeit a negative one" (p. 1). Even the "multiple modernities" literature establishes the West as the origin of multiple modernities (ibid., p. 67). Part one of the book calls for scholarship to go further in challenging the "continued privileging of the West as the universal 'maker' of universal history" by questioning the socio-historical evidence for ideas of rupture and difference and reconsidering the "conceptual framework of modernity from a wider spatial and historical context, *one which regards the very concept of modernity as problematic*" (ibid., pp. 1–2). Part two of the book examines and rewrites the central "myths" (European cultural integrity, the modern nation-state and industrial capitalism) that form the basis of social theories of modernity. The book concludes by offering a new approach—"connected historiographies"—for understanding the modern past that does not "fail to consider the histories of imperialism, colonialism, and slavery that enabled Europe, and the West, to achieve 'supremacy'" (ibid., p. 146).

Manfred B. Steger's (2008), *The Rise of the Global Imaginary: Political Ideologies from the French Revolution to the Global War on Terror*, puts an analytic spotlight on the emergent "new global imaginary" and has catalyzed a new generation of global studies as well as a renewed focus on the ideological dimensions of globalization. Citing transnational terrorist networks, climate change and global pandemics, Steger asserts that this emergent imaginary is "destabilizing taken-for-granted meanings and instantiations of the national" (p. ix), most significantly as they concern "the macrostructures of community and the microstructures of personhood" (p. 12). The book provides a useful delineation of the relationship between ideology and social imaginaries. Returning to its original "critical spirit of concept neutrality," Steger defines ideology as a "comprehensive belief systems composed of patterned ideas and claims to truth" (ibid., p. 4). Ideologies emerge within (and ultimately reinforce) "social imaginaries," which he defines as "prefigurative frameworks" constituting "the macromappings of social and political space through which we perceive, judge, and act in the world [and which provide] the most general parameters within which people imagine their communal existence" (ibid., p. 6). Though seemingly intangible, social imaginaries acquire materiality through the "(re)construction of social space and the repetitive performance of certain communal qualities and characteristics" (ibid., p. 7).

Much of this "work" is done via the deployment (and decontestation) of ideological concepts by ideological codifiers (typically the social elites of a society). In order to demonstrate this in practice, Steger identifies the core constructs and codifiers of five grand ideologies of the nationalist era—liberalism, conservativism, socialism, communism, Nazism/fascism—as they have diversely translated "the overarching national imaginary into concrete political doctrines, agendas and spatial arrangements" (ibid., p. 10). As the premise of his book implies, imaginaries are ultimately "temporary constellations" that can and have changed "with lightning speed and ferocity" (ibid., p. 7). Steger asserts that new imaginaries arise in moments of rupture ("the national" in the context of the French Revolution and European Enlightenment, "the global" in the context of the post-world wars and Cold War). Difficult as it is to describe an emergent process, Steger offers an insightful introduction to three emergent globalisms—market globalism, justice globalism and jihadist globalism—that are not yet full-scale ideologies but rather "ideological translators of the global" (ibid., p. 13). Key among his insights are that these seemingly distinct, if not divergent, social formations share deep-seated assumptions (about time and space) and, crucially, are still being articulated and contested.

Picking up four years later—and in the context of the global financial crisis and global justice movements of 2007/8—Manfred B. Steger and Paul James' (2013), "Levels of subjective globalization: Ideologies, imaginaries, ontologies," reiterates the "strange neglect" in the globalization research of the subjective meanings, ideas, sensibilities, and understandings, associated with globalization's "objective" processes (p. 19). Updating Steger's original articulation, they contend that the global imaginary is constituted by four conflicting and reinforcing ideologies: market globalism, justice globalism, religious globalism and imperial globalism, which form "a complex, roughly-woven but patterned, ideational fabric that increasingly figures the global as a defining condition of the present while still remaining entangled in the national" (ibid., p. 29). In order to understand the enduring dominance of "market globalism," the essay proposes an elaborated analytic framework intended to excavate the deeper layers of subjective globalization, namely the enduring and "profound ontological dominance of the modern" (ibid., p. 38) defined as "a (contingent) periodizing term, which names not the totality of a period within a particular spatial setting, but rather the uneven dominance of subjectivities and practices of 'the modern' within and across overlapping spatial settings" (ibid., p. 20). With a specific interest in "justice globalism" as the main ideological contender to "market globalism," the authors assert that until "ideologues" (used in a neutral way to refer to elite codifiers) address the taken-for-granted ontologies of modern time and space that underpin this general sense of the global, "they will remain confined to the dominant frameworks of the world they criticize" (ibid., p. 27).

Taking up the contemporary flourishing of social imaginary research, Suzi Adams et al.'s (2015) substantive introduction to the inaugural issue of *Social Imaginaries* presents social imaginaries both as an established field with well-defined intellectual sources, approaches and subjects as well as a "paradigm-in-the-making" marking "a qualitative shift in the way that social, cultural and political phenomena are understood and problematised"

(2015, p. 16). Their convocation of a new journal and collective (the Social Imaginaries Editorial Collective), and with it a generation of social imaginary research, is founded upon an understanding of imaginaries as "authentically *creative* (as opposed to imitative)" and as fundamental to "socio-political critique: for to be able to change social worlds, means that social worlds can be problematised and put into question" (ibid., pp. 42–43). The essay provides an extensive survey of the core and correlate sources, concepts and problematics explored under the umbrella of social imaginaries in three parts: theorizing social imaginaries, the history of imagination and social imaginaries and modernity.

The introduction reiterates the enduring significance of theorists Cornelius Castoriadis, Paul Ricoeur and Charles Taylor's theoretical frameworks for understanding social imaginaries. The essay also provides a historical journey of the concept of "imagination" from Aristotle and the Greek tradition, through Kant and the German Romantics and Idealists (with specific attention to the ideas of Martin Heidegger), with a brief mention of the significance of Sigmund Freud and John Dewey as well as its development in East Asia in the works of Miki Kiyoshi and Nakamura Yujiro, and finally back to Castoriadis. The essay highlights Castoriadis's notion of the "radical imagination" as the creative "magma" out of which societies are continuously reimagined. Taking radical in its etymological sense as "root," they assert that Castoriadis defined the radical imagination as the creative force (both on the level of the human psyche and the social whole) that creates "ex nihilo" that precludes on the one hand human's and human society's reduction to functionality and predetermination, and on the other hand creates the conditions for creativity, novelty and rupture. They with a discussion of contemporary formulations of modern imaginaries, including political-economic imaginaries, ecological imaginaries and multiple modernities as they diversly attempt to interpret the present.

Social Imaginaries produced ten issues from 2015 to 2019 that taken together highlight the productive heterogeneity and analytic utility of the social imaginary research in "identifying the existence of meaningful social practices and in elucidating movements towards social change" (Adams et al., 2015, p. 42). Essays in its inaugural volume variously presented and teased out foundational theorists and concepts demonstrating the strengths and potential of the field. Subsequent volumes took on the task of "making" the paradigm—working forward and backward—alternately opening the area to include thinkers not previously associated with social imaginary research and introducing emergent contributors, filling in geographical, cultural and conceptual gaps that crucially reinterpret the field from diverse geo-cultural perspectives, and introducing new terminologies and analytic frameworks.

Suzi Adams and Jeremy C.A. Smith's co-edited book, *Social Imaginaries: Critical Interventions*, distils the insights produced by their journal's Editorial Collective on a field they assert "continues to redraw, expand and problematise its boundaries" (2019, p. 35). The volume both maps the field's "idiosyncratic" intellectual terrain and charts new pathways, specifically surrounding emergent imaginaries (e.g. feminist, political, capitalist, ecological and constitutional). This is no simple project given the field's "exceedingly diverse and interdisciplinary" intellectual sources and trajectories and the fact that it has "breached the confines of academic discourse" and entered public discourse (ibid., p. 26). Preface contributor George Taylor describes the interpretative character of social imaginary research in terms of its horizontal (the diversity/breadth of approaches and imaginaries), vertical (the layers of social meaning made visible by such research), temporal (the ability to understanding imaginaries past, present, future and the related ability to use imaginaries as modes of critique and creativity) and, finally, ontological (describing social imaginaries as a core aspect of the human condition) dimensions. The Collective's greatest contribution was in providing an intellectual space to bring together and generate rigorous research on this far-ranging and timely concept.

Even so, Paul James' recent contribution to the 2019 edited volume, *Revisiting the Global Imaginary*, laments the continued imprecision in the term's usage suggesting this may stem from its lingering association with *zeitgeist* or *esprit du temps*, concepts that carry a cosmological/metaphysical connotation inconsistent with its contemporary usage. In an effort to break this problematic association, the chapter provides a genealogy of the concept (and its key thinkers) as it has been adapted, often uncritically, from its original cosmological sense to its current constructivist applications. Consistent with his corpus, this essay is a further attempt to bring conceptual clarity in the pursuit of analytic utility. In this vein, James reiterates Steger's original articulation of *ideas, ideologies* and *social imaginaries* as "an integrated set of levels of social engagement with meaning" (emphasis in original, p. 42) to which he and Steger later added *ontologies* (2013). He defines ontologies as foundational categories that form and frame human existence that are "historically constituted through the structures of human interrelations: temporality, spatiality, corporeality, epistemology and so on" (ibid., p. 43). The addition of "ontologies" crucially bridged their work with the "multiple modernities" literature and provided an essential layer of analysis for scholars and activists interested in social change. As James and Steger note in their 2013 essay about the dominance of market globalism over other global imaginaries:

> The global justice movement is yet to address the layers of subjective globalization in reflexive and systematic ways that shed light on the nature of the global imaginary, and, most markedly, the profound ontological dominance of the modern. It is this uncontested subjective and objective ground that gives neoliberal market globalism much of its strength. Unless Occupy and other global justice movements address these deeper subjective levels—and not only the objective practices associated with them—they will limit themselves to a mere expression of utopian hopes

> for the overthrow of what they do not like. Emancipatory practice, however, *requires global justice activists to set up the ideological, imaginary and ontological conditions indispensable for the creation of "another world."*
> (p. 38; emphasis added)

The radical imagination

The past several years point to a new generation of social imaginary research responsive to calls—like James and Steger's and Gibson-Graham's—for frameworks that inform emancipatory practice and that respond to the problem of how to imagine otherwise. While social imaginary research is inherently oriented toward the question of social change, research until now has largely been invested in weaving the field's diverse intellectual strands together in order to theorize (and make the case for) the creative sources of social formations and construct an analytic framework capable of describing and deconstructing modern social formations (in their temporal and spatial complexity and diversity). Max Haiven and Alex Khasnabish's (2014) *The Radical Imagination* is symbolic of shifts in social imaginary research and academia writ large that are reorienting academic labor and theory to the pursuit of cultivating the radical imagination and nurturing alternative economic practices, subjects and sites.

Haiven and Khasnabish (2014) describe the "imagination" both as a capacity to think about those things we do not or cannot directly experience as well as the filter or frame through which we interpret our own experiences. The social imagination is thus "something we do (and do together)" (p. 218). They offer three intertwined "tenses" through which we can understand the idea of the imagination: The first is as a conscious creative force of the individual mind (always influenced by and influencing others). The second is social as in "broad sets of landscapes of shared understandings and narratives that make living together possible" (ibid., p. 5). The third is the psychoanalytic/philosophical sense, which posits "the imaginary" "as a deep force at the very basis of the human subject … where ideas, meanings, associations, fixations, drives and affects circulate beneath the threshold of conscious thought" (ibid., p. 5). While ideologies are central components of social imaginaries—they are understood as a distinct conceptual category and as material, albeit transient, representations of "the imaginary relationship of individuals to their real conditions of existence" (Althusser, 1971). Ideologies are thus localized and historically situated attempts to organize, order and steer the "much more messy, complicated, contradictory and volatile" imagination (Haiven and Khasnabish, 2014, p. 230). As such, identification of core ideologies as they manifest in language and practice offer a means of de-construction and re-imagination.

Whereas Cornelius Castoriadis (1975) described the radical imagination it as the "magma" out of which all social formations and identities are constructed and the creative source and constant possibility of an/other social imaginary, Haiven and Khasnabish reinterpret it not as substance but rather as an *aspirational project and collective process* invested in doing the radical work of unearthing "deeply rooted tensions, contradictions, power imbalances, and forms of oppression and exploitation" (2014, p. 5) and imagining "the world, life and social institutions not as they are but as they might otherwise be" (2014, p. 3). To imagine "radically" is not to be contrarian or extreme, it is to engage in the collective and continuous process of assessing and addressing social needs.

According to Haiven and Khasnabish (2014), the radical imagination is ignited by the experience of difference and alienation, of the lingering sense that other things are valuable. Echoing Rebecca Solnit's (2004) important articulation of "hope in the dark," Haiven asserts that the "perversion, mutation and distortion of our hopes and dreams" is fundamental to the reigning system's functioning. As David Graeber puts it, "hopelessness isn't natural. It needs to be produced" (2008, p. 31). And while neoliberalism has had a stranglehold on how we imagine, express, act and reproduce what we value, Haiven and others see sparks of hope in the radicalization of the imagination in, as Susan George puts it the "hundred daily struggles" occurring all over the world in and against global capitalism (2004, p. 96). Angelos Varvarousis (2019) asserts that the most recent financial crisis was "catalytic" to the development of the radical imagination. Varvarousis describes crises as ambiguous moments of possible transformation that should be understood as *generative moments* as opposed to decisive points of "death or rebirth" (2019, p. 8, emphasis added). Though he is aware of the ways in which crises are often mobilized and abused by elites, Varvarousis contends that

> crises are important because they destabilize social imaginaries and open up a stage of suspension—a liminal stage—in which the rise of new social practices can facilitate the emergence of new social imaginary significations and institutions that can contribute to the alteration of the social imaginary at large.
>
> (ibid., p. 1)

Whether crises facilitate meaningful discussion and change, depends very much on the context, frameworks and ideas circulating when and where they occur. We contend that the contemporary discourses and practices of "commoning" and the "commons" hold incredible potential as tools and platforms for creating and embodying alternate social realities.

Commoning in, against and beyond capitalism

Many theorists have argued for the importance of "commons" in providing alternative spaces and forms of social cooperation "where we can reproduce ourselves and our world outside the dictates of the reigning paradigm" (Haiven, 2014, p. 252; see also Asara, 2020; De Angelis, 2017; Fishwick and

Kiersey, 2021; Gibson-Graham et al., 2013; Varvarousis, 2020). Contemporary approaches to commons prefer "commoning" as it shifts our attention from place to "a relational process—or more often a struggle—of negotiating access, use, benefit, care and responsibility" of property of any type, or any kind of communal wealth (Gibson-Graham, 2006b: p. 96). Describing the most recent iteration of commoning movements and practices—public square "occupations"—Haiven asserts that commons "act as laboratories of the radical imagination, places to imagine collectively and put into practice new social relations and new forms of cooperation" (2014, p. 74).

Many intellectuals describe the post-2008 proliferation of public square movements (e.g. los Indignados, Occupy and the Arab Spring) as "manifestations against enclosure" that are literally and symbolically forcing political ideas, social issues and economic relations back onto the public stage and retaking the commons (Haiven, 2014, p. 75; see also Appel, 2014; Asara, 2020; Castells, 2017; Harvey, 2012; Varvarousis, 2019; Vogelaar, 2015). Haiven's description of commons as "zones for reimagining social, political and economic relations" draws attention to the relevance of "prefigurative politics" in the radical imagination (ibid., p. 74). Prefigurative politics refers to a "general shift in emphasis from attempts to seize the state apparatus or influence existing socio-political systems and towards the construction of alternative futures in line with the aspirations animating social justice struggles" (Haiven and Khasnabish, 2014, p. 62). The politics of prefiguration can be traced to the so-called "new social movements" of the 1960s—peace, queer, anti-colonial, civil rights, student, feminist and environmental—that emerged in the wake of World War II and sought to expand activism beyond the political and economic realm by focusing upon the ways in which social formations are embodied and re/produced in large part through our daily activities. According to McKay, the history of twentieth-century social justice struggles should be read as a series of "experiments in living otherwise" (2005, p. 174) geared towards building alternative practices, social structures, communities, and identities. The countercultural movements of this period saw the creation and expansion of various prefigurative projects that live into this day including: housing squats and cooperatives, alternative educational initiatives, direct action collectives, Indy and alternative media organizations, makerspaces, Indigenous and migrant solidarity groups, sustainable transportation initiatives, consciousness raising initiatives and groups, and various other forms of radical being and belonging (Haiven and Khasnabish, 2014). Far from being mere acts of utopian theater, these prefigurative projects aimed at bringing "possible futures 'back' to work on the present, to inspire action and new forms of solidarity today" (Haiven and Khasnabish, 2014, p. 3).

In addition to providing the common grounds for social change, commons are a central medium through which other-than-capitalist subjectivities can be prefigured and cultivated. As Gibson-Graham (2006b) they assert, an expanded conception of economy hinges on an expanded range of economic language, practices, subjectivities and communities. Indeed, alongside language, the subject is a crucial site for social change insomuch as its materialization is the product of the continuous repetition of ritualized practices and interactions that can be subverted and changed (Gibson-Graham, 2006b). As a form of capitalistic activity that takes place in large part on the level of the subjective self, labor constitutes an important source for social transformation.

Reimagining labor/labor as a commons

Reimagining labor is an important but difficult step in reimagining our economic systems, for (waged) work constitutes a central (and for the most part unquestioned) feature of our lives: it is how we spend most of our waking time, it is a source of income and a means of self-accomplishment, social recognition and socialization, and for many, it is a constitutive feature of one's identity (Gheaus and Herzog, 2016; Livingston, 2016). While work can be meaningful, it can also be a source of abuse and exploitation, alienation and existential dread, and increasingly, of "corporal rift" (Foster and Clark, 2020). What is more, work conditions differ widely depending on the field and context, and while some work is well-respected and renumerated, other work is not even considered work (Gibson-Graham, Cameron and Healey, 2013). As degrowth scholar Timothée Parrique so succinctly puts it, "work is the problem but it is also part of the solution" (2019, p. 566).

Inspired by the contemporary resurgence of commoning movements and practices across the world and informed by the critical literatures on social reproduction and work in the capitalist imagination, our project examines practices and processes of recommoning labor in three community projects. Building upon scholar-activist Tommaso Fattori's (2011) expansive conception of the commons, sociologist and feminist activist Hilary Wainwright encourages us to break with mainstream understandings of labor as "profoundly individual" (2014, p. 76) and to reimagine labor as a commons. According to the author,

> by naming this creative capacity, this characteristic of all of humanity, as a commons, and highlighting its social as well as individual character and the associative, social conditions of its realization, we also lay the basis for reclaiming the products of this capacity.
>
> (ibid., p. 76)

Using the example of "Creative Commons" licensing, she contends that commoning approaches are well suited to negotiating the tensions between the "collaborative dimension of creativity and the varying necessity for individual autonomy, introversion and self-reflexivity" (ibid., p. 78). Describing her own experience working in a "solidarity economy media enterprise," *Red Pepper*, Wainwright notes that organizations committed to embodying labor as a commons must recognize:

> diverse sources of support, monetary and in kind, some from organizations and some from individuals, all of whom expect some accountability. It also has to recognise several sources of creativity, the importance of collaborative editorial process, the dimension of individual decision-making at different levels of the project, and the need for relatively coherent identity. The notion of creativity as commons seems key to developing a sufficiently flexible, transparent and constantly negotiable form of governance to deal with this complex combination of interests and imperatives.
>
> (p. 78)

In addition to reimagining productive labor as a commons, we must also reimagine reproductive labor along common lines. As O'Connor suggest, the reproductive crisis we find ourselves in requires that we form new social alliances and new forms of worker solidarity that are forged

> along lines that are "communal" as much as industrial, extending not just over workplace solidarities on such matters as occupational health and safety, working conditions, hours of employment and wage levels, but also to the maintenance and enhancement of collective livelihood—including such matters as toxic waste control, respect of habitat as a cultural as well as ecological milieu, recognition of importance of town and village solidarities, friendship, assistance, etc.
>
> (1994, p. 5)

Radical research in the neoliberal university

While universities have always been tied to the production and reproduction of the social orders in which they are enmeshed, they are also central sites of struggle (Haiven, 2014). Though we are accustomed to thinking about the university as an "ivory tower" separate from, indeed held above, the rest of society, universities have become central to global capitalism and a "key example of the struggle over values and imagination that animate our historical moment" (ibid., p. 135). According to Haiven,

> what has happened to universities is not unlike what is happening to practically every other sector of the economy under neoliberal austerity capitalism: privatization of public institutions, the "rationalization" of workforces, the increase in part-time and precarious labour, the commodification of knowledge and time as a "service" to be bought and sold, and the imposition of market-derived measurements and forms of discipline.
>
> (ibid., p. 135)

Neoliberal changes to the structure and management of universities have altered university missions, staffing and "output" favoring market-friendly research and researchers and professional and high-tech programs that "service monopoly capitalism" (Slaughter and Rhoades, 2000, p. 75). These shifts have peripheralized the civic and creative functions of the university and seen the correlate marginalization of the humanities and liberal arts. They have also seen the "enclosure of knowledge, where disciplinary boundaries, increasingly corporatized research and commercialized spaces shape and constrain human possibility" (Haiven, 2014, p. 22).

Beyond engaging in occupational navel-gazing, Haiven and Khasnabish assert that "the politics of the university *matter* in an age of 'cognitive capitalism'" (2014, p. 39) when intellectual labor is increasingly mobilized to serve, not the social good, but rather the market. Like the factory of the industrial age, the university has become a central laboratory for developing new ways to discipline labor (Haiven, 2014). This occurs on the level of the student whose education is increasingly oriented to "job-ready skills" and funded through their own debt and whose sense of self and of possibility are "finely calibrated to the capitalist values of accumulation, competition and individualism" (ibid., p. 149). It also occurs at the level of employment as universities are increasingly run on a reserve army of surplus labor and as research and teaching are increasingly oriented to, and funded by, economic imperatives. Survival in the contemporary university demands continuous rebranding around a persistently shifting set of market demands. In our experience, this does not bode well for the quality of instruction or research.

Even as they have been co-opted by the neoliberal values of growth and profit, universities remain important sites of struggle for several reasons: First and foremost, the volume of people who pass through universities (as employees and students) makes the university, like the factory of the industrial era, a potential site for mass resistance and radical reimagination. What is more, as places "where our sense of value and our imaginations are honed and tailored" (ibid., p. 136) and new knowledge is produced and disseminated, the university still has the potential to be "a space–however imperfect—of critical and free inquiry" (Haiven and Khasnabish, 2014, p. 36). As academics who trained and served in—and have been recently discarded by—the neoliberal university, we take both the threats to, and potential of, the university seriously and see this research project as a part of a broader project to convoke the radical imagination and re-common education through our teaching and research.

Our project borrows and builds upon Haiven and Khasnabish's (2014) and Gibson-Graham's (2006a, 2006b) action research projects as they use academic labor as a means of providing "the opportunities, resources, time and space necessary to collectively bring into being the prefigurative capacity to envision and work towards building better worlds" (Haiven and Khasnabish, 2014, p. 61). Arguing for its radical potential, Haiven and Khasnabish assert that research is one important way that communities reproduce themselves and transform their wider social contexts. They propose "convocation" as an aspirational (prefigurative) form of research that seeks to awaken, sharpen and enliven movements' inherent capacities for "research" in the broader sense (ibid., p. 2). As an approach to social movement research, convocation calls for "a reflexive and responsive relationship between the researcher and the movement(s)" and a reflexive account of the assumptions that form and frame the "research imagination" (ibid., p. 216). Far from being a means of acquiring "academic capital," solidarity research of the sort promoted by Haiven and Khasnabish and Gibson-Graham seeks to "break those sedimented silences of history and [interrupt] the monotonous normative reproduction of social life and its accumulated injustices" (ibid., p. 56) by harnessing "social science as a crucial organ or circuit of the constant work of community reproduction" (ibid., p. 254).

Two of the communities studied in this book were a part of academic travel courses we taught as a part of our department curricula.[3] We were drawn to these communities as manifestations of experiments in living and working otherwise. Our subsequent discussions about these visits inspired us to think about their common and diverse approaches to small-scale, everyday world building, and to write this book.

Case studies in re-commoning labor

In order to participate in the broader project of convoking the radical imagination, we focus on three geographically, historically and strategically diverse community economies/projects. Our aim is to understand each community's approach to reimagining labor through an analysis of their material practices, discourses and self-reflection. In order to explore emergent approaches to re-commoning labor we studied communities in the US American Southwest (Tonatierra), in the foothills of the Indian Himalayas (Navdayna) and in the Highlands of Scotland (Tombreck). The rationale for these selections is fourfold: First, we chose projects from different geographies, histories, practices and people as they offer a view of situated and context-specific "commoning" in and against neoliberal capitalism. As Otero has stated, subaltern actors "in different parts of the world or in different regions within a country ... may have diverse structural capacities depending on their distinctive histories and cultures, or the villages or regions where they develop" (1999, p. 22). Second, and relatedly, each community is (diversely) embedded in colonial, decolonial, and neocolonial contexts and histories that, taken together, help us reflect upon the variegated influences and residues of colonialism in economic relations and forms. Third, each project came into existence in the historical context of the various anti-capital, anti-colonial global justice struggles of the 1990s and is animated by their language, ideas and practices. And finally, each community constitutes a different (and hopeful) exemplar of the "routine, banal and often heart-wearying labour" of social transformation (Haiven and Khasnabish, 2014, p. 20). As Gearey and Ravenscroft (2019) assert, in focusing only on archetypical agents of activist intervention we miss the "plethora of ordinary, pedestrian, unrecognized nowtopian practices burgeoning in unrecognized corners … generating environmentally sensitive ways of being which exist in the everyday" (p. 25).

Methodology

In order to study these three sites of dynamic social imagination, we employed interview, survey and discursive methods. Research was conducted between summer 2020 and spring 2022. Site visits of Tombreck and Navdanya in 2018 and 2019 inspired our early research interests in the communities and informed our approaches. Our original study plan involved visits to all three sites in 2020 and 2021, but these plans were unfortunately curtailed by Covid-19. The following chapters present the results of the first part of what we hope is a much longer collaboration with these research sites and their community members. Given Covid-related travel limitations, this book focuses on the insights gained from analysis of textual materials (websites, documents, manifestos, maps, etc.) and interviews. The first stage of research involved content analysis (both of digital materials and interviews) to uncover key themes, terminologies, issues and values. These findings were used to develop better questions for follow up interviews and content analysis. The findings of these will be used in turn to help us develop better questions and strategies for the next stages of collaborative research.

Notes

1 As opposed to psychoanalytic theories of the individual imagination as in Jacques Lacan's notion of the Imaginary as one of three dimensions of identity formation (Imaginary, Symbolic, Real) or Jean-Paul Sartre's phenomenological account of the existence and role of the imagination in human consciousness.
2 Not to be confused with Karl Marx's complete insights, which were decidedly complex and varied and naturally changed over time.
3 See Dasgupta (2021) for a detailed discussion on how Navdanya was used as a site to teach students about South-North knowledge transfer.

References

Adams, S. Blokker, P., Doyle, N.J., and Krummel, J. (2015) Social imaginaries in debate. *Social Imaginaries*, 1(1), pp. 15–52.

Adams, S. and Smith, J.C. (eds.) (2019) *Social imaginaries: Critical interventions*. Lanham: Rowman & Littlefield.

Althusser, L. (1971) Ideology and ideological state apparatuses. Available at: www.marxists.org/reference/archive/althusser/1970/ideology.htm.

Anderson, B. (1983) 1991. *Imagined communities: Reflections on the origin and spread of nationalism*. London: Verso.

Appadurai, A. (1990) Disjuncture and difference in the global cultural economy 1990. *Theory, Culture & Society*, 7(2–3), pp. 295–310.

Appadurai, A. (1996) *Modernity at large: Cultural dimensions of globalization*. Minneapolis: University of Minnesota Press.

Appel, H. (2014) Occupy Wall Street and the economic imagination. *Cultural Anthropology*, 29(4), pp. 602–625.

Arnason, J.P. (2015) Theorizing capitalism: Classical foundations and contemporary innovations. *European Journal of Social Theory*, 18(4), pp. 351–367.

Asara, V. (2020) Untangling the radical imaginaries of the Indignados' movement: Commons, autonomy and ecologism. *Environmental Politics*, pp. 1–25. doi: 10.1080/09644016.2020.1773176

Baczko, B. (1984) *Les imaginaires sociaux: mémoires et espoirs collectifs*. Paris: Payot.

Bhambra, G.K. (2007) *Rethinking modernity: Postcolonialism and the sociological imagination*. New York: Springer.

Castells, M. (2017). *Another economy is possible: Culture and economy in a time of crisis.* Cambridge: Polity.

Castoriadis, C. (1975) *The imaginary institution of society*. Cambridge: MIT Press.

Dasgupta, P. (2021) Using academic travel to teach sustainable economic development. *Wirtschaft neu lehren, Sozioökonomische Bildung und Wissenschaft*, pp. 125–143. doi.org/10.1007/978-3-658-30920-6_9

DeAngelis, M. (2017). *Omnia sunt communia: On the commons and the transformation to postcapitalism*. London: Zed Books.

Fattori, T. (2011) Fluid democracy: The Italian water revolution. *Transform! Magazine*, 9.

Fishwick, A. and Kiersey, N. (2021) *Postcapitalist future: Political economy beyond crisis and hope*. London: Pluto Press.

Foster, J.B. and Clark, B. (2020) *The robbery of nature: Capitalism and the ecological rift*. New York: NYU Press.

Gaonkar, D.P. (2001) *Alternative modernities*. Durham: Duke University Press.

Gaonkar, D.P. (2002) Toward new imaginaries: An introduction. *Public Culture*, 14(1), pp. 1–19.

Garner, J.V. (nd) The break from Marxism. In Cornelius Castoriadis (1922–1997), *Internet encyclopedia of philosophy*. Available at: https://iep.utm.edu/cornelius-castoriadis/#H2

Gearey, M. and Ravenscroft, N. (2019) The nowtopia of the riverbank: Elder environmental activism, *Environment and Planning E: Nature and Space*, 2(3), pp. 451–464. doi: 10.1177/2514848619843733

George, S. (2004) *Another world is possible if* … London: Verso.

Gheaus, A. and Herzog, L. (2016) The goods of work (other than money!). *Journal of Social Philosophy*, 47(1), pp. 70–89.

Gibson-Graham, J.K., Resnick, S. and Wolff, R. (2001) Toward a poststructuralist political economy. In Gibson-Graham, J.K., Resnick, S. and Wolff, R. (eds.) *Re/presenting class: Essays in postmodern Marxism*, Durham, NC: Duke University Press, pp. 1–22.

Gibson-Graham, J.K. (2006a) *The end of capitalism (as we knew it): A feminist critique of political economy*. Minneapolis: University of Minnesota Press

Gibson-Graham, J.K. (2006b) *A postcapitalist politics*. Minneapolis: University of Minnesota Press.

Gibson-Graham, J.K., Cameron, J. and Healy, S. (2013) *Take back the economy: An ethical guide for transforming our communities*. Minneapolis: University of Minnesota Press.

Graeber, D. (2008) Hope in common. *The anarchist library*. Available at: https://theanarchistlibrary.org/library/david-graeber-hope-in-common

Haiven, M. (2014) *Crises of imagination, crises of power: Capitalism, creativity and the commons*. London: Zed Books.

Haiven, M. and Khasnabish, D.A. (2014) *The radical imagination: Social movement research in the age of austerity*. London: Bloomsbury Publishing.

Harvey, D. (2012) *Rebel cities: From the right to the city to the urban revolution*. New York: Verso books.

James, P. (2019) The social imaginary in theory and practice. In Hudson, K. and Wilson, E.K. (eds.) *Revisiting the global imaginary*. Cham: Palgrave Macmillan, pp. 33–47.

Livingston, J. (2016) Fuck work. Aeon. Retrieved on June 2, 2022 from https://aeon.co/essays/what-if-jobs-are-not-the-solution-but-the-problem

McKay, I. (2005) *Rebels, reds, radicals: Rethinking Canada's left history*. Toronto: Between the Lines.

Marx, K. (1887) *Capital, Volume I*. Moscow: Progress Publishers.

O'Connor, M. (1994) The second contradiction of capitalism: The material/communal conditions of life. *Captialism, Nature, Socialism*, 5(4), pp. 105–114. doi: 10.1080/10455759409358613

Otero, G. (1999) *Farewell to the Peasantry? Political class formation in rural Mexico*. Boulder: Westview Press.

Parrique, T. (2019) *The political economy of degrowth* (Doctoral dissertation, Université Clermont Auvergne (2017–2020)).

Solnit, R. (2004) *Hope in the dark: Untold histories, wild possibilities*. Chicago: Haymarket.

Slaughter, S. and Rhoades, G. (2000) The neo-liberal university. *New Labor Forum*, 6(Spring–Summer), pp. 73–79.

Steger, M.B. (2008) *The rise of the global imaginary: Political ideologies from the French revolution to the global war on terror*. Oxford: OUP Oxford.

Steger, M.B. and James, P. (2013) Levels of subjective globalization: Ideologies, imaginaries, ontologies. *Perspectives on Global Development and Technology*, 12(1–2), pp. 17–40.

Taylor, C. (1999) Two theories of modernity. *Public Culture*, 11, pp. 153–174.

Taylor, C. (2004) *Modern social imaginaries*. New York: Planet Books.

Varvarousis, A. (2019) Crisis, liminality and the decolonization of the social imaginary. *Environment and Planning E: Nature and Space*, 2(3), pp. 493–512.

Vogelaar, A. (2015) *Staging revolution: The OWS encampment at Zuccotti Park*. Santa Barbara: Media Fields.

Wainwright, H. (2014) Notes for a political economy of creativity and solidarity. In Satgar, V. (ed.) *The solidarity economy alternative: Emerging theory and practice*, Durban: University KwaZulu Natal Press, pp. 74–79.

4 Plotting, planting and poesies[1]

Conjuring the commons at Tombreck Farm, Scotland[2]

Introduction

Tombreck Farm is a rural regenerative farm and residential community in Highland Perthshire, Scotland, set on 240 acres of land running from the northern banks of Loch Tay to the southern slopes of Ben Lawyers.[3] Highland Perthshire occupies a central place in Scottish geography, history and culture. The region is due north of Scotland's urban, central belt, and it is divided by the Highland Boundary Fault—a geological and cultural fault line. It is home to Scone Palace, the ancient residence of Scottish monarchs, and its city of Perth was once a capital of Scotland. Scotland's "image maker," Sir Walter Scott, proclaimed, "Perthshire forms the fairest portion of the northern Kingdom" (1871, p. 15).[4] Owing to its geology, Perthshire is a landscape of contrasts: with a rugged, mountainous area in the north, a pastoral, agricultural area in the south, and the country's longest river, the Tay, running through its center. The region is often referred to as "big tree country" and has 200,000 acres of woodland, including one of the oldest living trees in Europe—the Fortingall Yew Tree, which is estimated to be between 3,000 and 5,000 years old (Bevan-Jones, 2004).

Situated in northwest Perthshire, Tombreck Farm encompasses a diverse landscape of woodlands, wetlands and grasslands. The woodlands are a part of the larger Carie/Cragganester Woods, which have been designated a "Site of Special Scientific Interest."[5] Because it is southern facing and amenable to cultivation and farming, there is a long record of human activity on the land; indeed the ruins of several old settlements can be found scattered throughout the property. The current landholder, Tober Brown, inherited the property from his father in 1997 at which point much of the land had been neglected and over-grazed and the buildings and infrastructure were rundown. Brown's vision for regeneration was inspired by a question—"how could a farm, which supported several families 150 years ago and now failed to support even one, be vibrant and productive once more?"[6]—and a promise—that he would make his mother proud. Over the past 25 years,

DOI: 10.4324/9781003138617-4

Brown, along with family and community members, has begun to regenerate the land, infrastructure and community. Tombreck presently has 20 residents, ten[7] residences, two community buildings, various utility sheds, a woodland agricultural system, a small flock of Castlemilk Moorit sheep, a pedigree herd of free-range KuneKune pigs, as well as a number of small land-based enterprises being operated on or from the site (e.g. including willow growing, Scottish native wildflower growing, and a market garden) (Tombreck, 2017).

As with each project examined in this book, Tombreck is informed and inspired by its colonial and neoliberal history and the varied cultural and political approaches to reviving and restoring Scotland's real and imagined past. We read Tombreck alongside many contemporary approaches to regeneration[8] in Scotland that are intentionally experimenting with different ways to work with and live on the land that borrow thoughtfully from lost traditions.

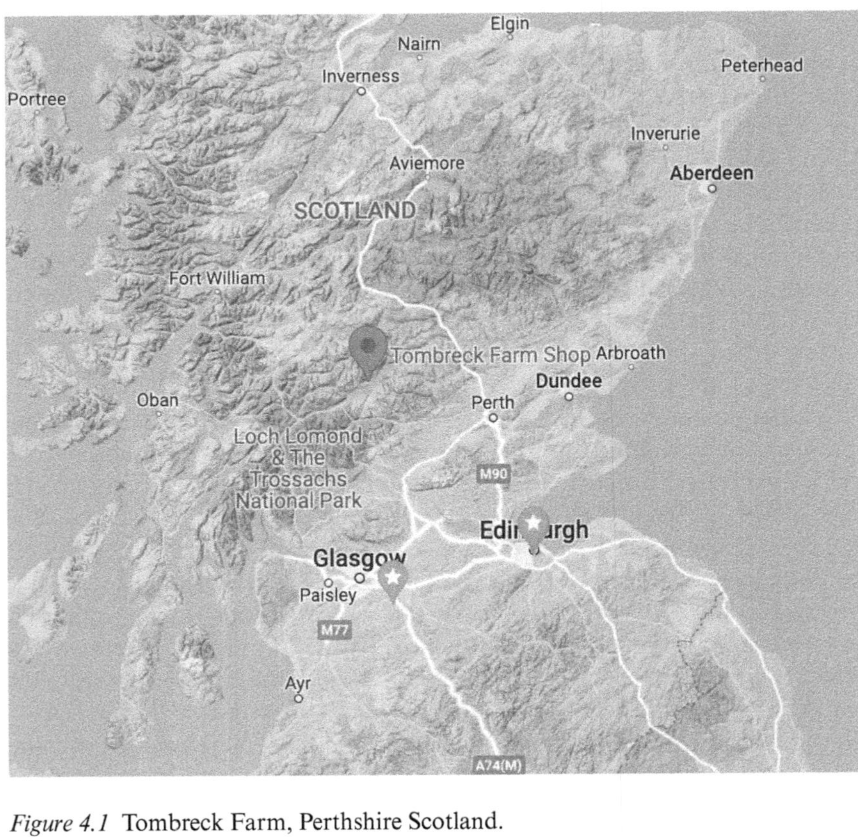

Figure 4.1 Tombreck Farm, Perthshire Scotland.

From time immemorial[9]

Tombreck sits on what was once the "Great Caledonian Forest," an old-growth temperate rain forest that covered much of Scotland and the British Isles following the last Ice Age (7,000–8,000 years ago). While the forest had been reduced significantly due to climate changes by 2000 BCE, it was all but decimated by the late nineteenth century due to human actions. Only 84 remnants of the forest remain (about 1% of the original) and Tombreck Farm sits on one of the largest in Perthshire (Todaro, 2022). There is abundant archaeological evidence of human habitation dating back to the Bronze Age in the area and the site is enmeshed in the rich clan history of Highland Scotland. Prior to the fourteenth century the site stood on the vast lands of Clan MacMillan and later in that century became part of the ancestral lands of the Campbells of Breadalbane whose influence stretched from their seat at the nearby Taymouth Castle in Kenmore to the Argyll islands on the west coast (Campbell, 2000). According to records, generations of Campbell lairds showed great interest in the woods; for example, Black Duncan Campbell, the seventh Laird of Glenorchy from 1583 until 1631, was a renowned planter and protector of trees who fenced off Tombreck and other lochside woods to protect them from deer and pasture animals (Tombreck, 2017). As the value of oak for fuel and building increased during the eighteenth century his successors enclosed many of the woods with impressive stone walls, the remnants of which can still be seen at Tombreck (Tombreck, 2017).

The property has been the site of a farm since at least the eighteenth century when it was a part of the land holdings of the family of Breadalbane who owned thousands of acres in the surrounding area and rented much of the land to tenant farmers and crofters.[10] Remnants of limekilns are evidence of fertilizer production for use in farming. Records indicate that the area was well populated at that time and there is no doubt that the woodlands would have been an integral part of people's livelihood, providing building materials, tools, fuel and food. Census records from 1841 indicate that Tombreck was divided into two farm holdings (Easter and Wester Tombreck), which were home to at least three households (extended families and servants) with a total of 21 inhabitants (Tombreck, 2017). By 1851, the land held two households, the McLarens and the Robertsons, with a total of 18 people (Tombreck, 2017). Between 1871 and 1901 only the McLaren's and their servants remained on the site. The census of 1901 records the McDiarmids living on the land; all eight residents are described as working at home (Tombreck, 2017). After the break-up and dispersal of the Breadalbane Estate in 1920, the farms that had previously been tenanted were sold (Tombreck, 2017).

In 1948, Tombreck was bought by father and son, Andrew and James Brown, who farmed the land for 35 years as a mixed stock farm; growing local crops of oats, turnips, kale, rape, and grass and raising cattle, sheep, lambs, and hens. Due to declining health, the Browns sold their livestock and rented the land out for grazing in the 1980s. From 1992 to 2017, the land was leased in a grazing agreement to Gordon Stewart who raised his family on the property and remains there today. In 1997, James and Ann Brown's son, Tober, inherited the land at which time he began the slow process of sustainably regenerating the farm: repairing drainage systems, cultivating the gardens, restoring the woodlands, reducing the number of grazing animals, repairing existing buildings and building new ones.

Tombreck today

Tombreck Farm presently has 20 residents (couples, families and single people) living in nine homes (the Farmhouse, Granny's House, the Dragon's Lair, the Straw Bale House, West Byre, East Byre, Yurt 1, Yurt 2 and the Manse), two community buildings (the Cart Shed Studio, an office and Farm Shop, and the Big Shed, a community building with a multipurpose hall, commercial kitchen, studio and workshop), a small flock of Castlemilk Moorit sheep, and a pedigree herd of free-range KuneKune pigs, as well as three horses and two goats. Many residents run small businesses from and/or on the site (e.g. an architecture firm, consulting work, market garden) and many also have employment offsite (e.g. local post office, museum). Some residents do not undertake waged work from the property but contribute to the maintenance of the shared services. Most residents have multiple jobs. Several local community members work from but do not live on the farm (e.g. bee keeping). The Farm Shop and the Big Shed are also used, rented and/or accessed by the general public.

The farm is owned by Tober Brown[11] and is managed in partnership with a board of directors constituted by family members.[12] The finances and day-to-day operations are overseen by Sue Manning and Tober Brown, who both live and work on the farm. The farm business is aided by the assistance of WWOOFers.[13] Manning is a registered architect and a founding partner in the all-female architecture firm Ecological Architecture, which she operates from the farm. Her sustainable design expertise and commitment to creating sustainable and affordable housing "into perpetuity" have been central to regenerating Tombreck since its beginnings (Manning, 2022). She is central to all renovation and new building projects. Manning is also responsible for Tombreck's designation as a ScotLAND Permaculture Centre[14] and co-founded and directs the Lochtayside Community Interest Company, which owns and runs the Big Shed community center at Tombreck.

All residents and tenants are members of the Tombreck Action Group, founded by Brown in 2003 as the organizational unit that would help guide and facilitate proposals, plans and implementation. The initial aim put forth by the Group was: "To work as a farm, and to diversify into other activities so as to include other people, and to provide housing and employment at a sustainable level" (Tombreck, 2017). Everyone living and/or working at Tombreck is required to be a member and to pay a monthly fee (presently

£10), which goes towards maintenance (e.g. sewage treatment system, water supply, the access track, the courtyard). Meetings are held twice a year and as needed.

The ethos of Tombreck is best described as a commitment to "communal self-reliance"[15] connected to the land, landscape and history (Tombreck, 2017). According to a 2022 document created for prospective applicants for two new schemes—the market garden and grazing tenancy:

> The long-term vision for Tombreck has always been to build a community of people who live and work on the farm and in the local area. Access to affordable housing is key to this vision. We want the farm to be managed regeneratively and productively providing a local, low carbon, resilient food production system. One that builds soil health, fosters biodiversity and protects water sources. The current aging demographic of the community is such that we are especially welcoming applications from young people and families.
>
> (Tombreck, 2022)

In 2003, the Tombreck Action Group created an "action plan" covering five areas including: land use (involving woodland regeneration and new planting as well as seed and soil cultivation), employment/development (involving job creation), existing buildings (renovations and repurposing of existing buildings), housing (involving three grant-funded upgrades and three new builds as well as the proposal for a housing association or cooperative) and infrastructure (involving water drainage, capture and disposal, fencing and tracks and electric and renewable energy. One interviewee noted that the plan changed a fair bit over the years as things developed and there were setbacks along the way. In addition to the longer-term goal of developing a housing cooperative,[16] the Action Group has created two new initiatives that build on the regeneration efforts described above and form an essential part of making Tombreck a serious farm business. One is for the development of a field-scale market garden and another is for a grazing tenant enterprise.

The early stages of regeneration outlined here were made possible in large part through several grant schemes including a Scottish Forestry Grant, an ESA (Environmentally Sensitive Area) grant, Housing Grants from the Perth and Kinross Council and the Communities Scotland Rural Empty Properties Grant. A new Empty Properties Grant has just recently been announced but grants are increasingly hard to come by due to changes in political priorities (e.g. the big grants that used to be available for community buyouts in rural areas are now earmarked for properties in urban areas) and, more significantly, the loss of EU grants that were formerly distributed by the UK government but which are now gone because of Brexit (Graham, 2022b). The above grant schemes are evocative of a larger effort in Scotland, beginning in the late 1990s under Scotland's newly formed parliament, to revive and regenerate Highland and Island communities, landscapes, and ecosystems. These efforts are a direct response to the dispossession of rural communities initiated during the eighteenth-century clearing and enclosure of Highland Scotland (to make way for "agricultural improvements" and absorb the region more fully into Great Britain) and reinforced centuries later by British Prime Minister Margaret Thatcher's neoliberal austerity programs.

The following section situates Tombreck's community development and natural regeneration programs in the broader colonial and neoliberal history of dispossession in Scotland as well as the culturally-specific responses to that dispossession on and from which Tombreck builds and borrows.

The Highland Clearances and the Thatcher years

> Ecologically, there is little left of Scotland. Lanced of danger, fully domesticated, the countryside has been kitted out as an English larder, a table laid with lamb and strawberries and clotted cream.
>
> (Fraser, 2010, n.p.)

American author Caroline Fraser's depiction above of a Scotland both tamed[17] and dispossessed, gestures at the persistent entanglement of Scottish politics, identity and landscape and of its complex ties with its neighbor to the south. Recent years have seen a spate of efforts to "rewild,"[18] "reforest,"[19] and "regenerate"[20] Scotland's natural and rural areas, efforts we believe can be read against the broader backdrop of Scotland's diverse and uneven experiences of, and reactions to, "internal colonialism"—a term Michael Hechter (1975) has used to describe English treatment of the Celts in parts of Ireland and Scotland and that has been subsequently applied to colonial contexts across the globe. According to Hechter, the Celtic regions of Ireland and Scotland were colonized by the English in much the same way as their overseas colonies: Celtic (Gaelic) regions and peoples were regarded as socially and culturally backward ("savage"), were politically dominated and economically exploited for labor and natural resources and forcibly Anglicized through religion, language and education (1975, 2021). While we recognize there is danger in oversimplifying the relationship between Scotland and England and underplaying internal divides and differences,[21] it is safe to say that Scotland occupies a peculiar place in the United Kingdom as a stateless nation and postcolonial site. In addition to providing fodder for thinking about the enduring (and often invisible) effects of colonialism, Scotland's postcolonial experience highlights the centrality of land as a commons—not only in the traditional sense as a shared natural resource—but also in the more dynamic sense as a source of cultural identity and ancestral ties.

The "Highland Clearances" refers broadly to a series of policies and programs implemented in the Highlands (a geological,[22] geographical and cultural designation) of Scotland from the mid-eighteenth to the mid-nineteenth centuries that were inspired by the Agricultural Revolution sweeping Great Britain at that time as well as the "Jacobite Uprisings" in the Highlands. In addition to clearing the Highlands to make way for sheep and "agricultural improvements," the Highland Clearances also sought to clear the Highlands of its (rebellious) people. The Jacobite Uprisings or Rebellions were a series of violent uprisings between 1688 and 1788 by the Jacobites (supporters of King James the VII of Scotland and II of England who sought to restore the Stuart dynasty to the throne). Though the uprisings were wrapped up in the centuries' long struggle for Scottish independence from England, the Jacobite cause was not straightforwardly about independence.[23] And while it is important to note that the Jacobite cause was motivated by a variety of political and religious issues that were internally divisive, and that many Highlanders were not Jacobites, it was nevertheless treated by the English crown as justification for the indiscriminate eradication of (predominantly Gaelic) Highland and Island communities and structures. The rapid and radical changes to Highland structures and economies brought on by the Clearances fractured (and eventually decimated) the clan system and its related organizational and agricultural structures and saw the poverty-induced mass migration of Highlanders to the Isles, Lowlands and abroad.[24] What is more, the Clearances dramatically altered Highland nature and ecosystems. As author Susan Wright et al. have put it, Scotland's iconic landscapes are "surrounded mostly by ecological deserts" (2018).

While the Clearances fundamentally transformed the social, political, economic, geographic, and ecological landscape of the Highlands and Islands of Scotland, they also inspired an immediate and abiding (as well as politically and economically potent) nostalgia for Highland lifestyles and symbols (Devine, 1983). Post-Clearance Highland revivalism is associated with the literary works of Robert Burns and Sir Walter Scott whose poetry, prose and "media events"[25] diversely romanticized and re-invented Highland life, landscape, and heroes and set in motion an enduring appetite (both within Scotland and abroad) for Highland regalia. Indeed, the Highlands (real and imagined) continue to play an important albeit contentious role in Scottish identity, economics and politics and to what Scottish activist and politician Jimmy Reid refers to as the "Scottish Question"—the seemingly intractable and "unfinished business" of Scottish independence (McCrone, 2001, see also Mitchell, 2014).

While the struggle for political independence is a recurring theme in Scottish history beginning with the formation of the Kingdom of Scotland in 843 AD (and prominently fought for in the Wars of Independence and Jacobite Uprisings), the first official "Home Rule" movement was taken up in 1853 in the wake of the Highland Clearances. The various attempts[26] to establish sovereignty (or some form of devolved powers) sputtered until discovery of North Sea oil in the 1970s made the independence movement and party (the Scottish Nationalist Party) viable, at least on economic grounds. The first contemporary referendum (for a devolved deliberative assembly) was held in 1979 against the backdrop of political infighting. While the referendum passed with 52%, it was repealed because a condition of the referendum (set by Westminster) required that 40% of the total electorate should vote in favor in order to make it valid. But because the turnout was only 64%, only 33% of the electorate voted in favor. Following the political battles surrounding the failed referendum, the SNP withdrew support from the Labour government in protest, filing a vote of "no confidence." The vote forced a general election in 1979, which was won by the Conservative Margaret Thatcher.

Thatcher came to office in the middle of an economic recession and a (post-imperial) identity crisis in the UK. According to Wood (2019), Thatcher's political tactics were informed by her refusal to accept "Britain's diminishment" and her resentment of the popular notion of Britain as a "poor nation whose greatness only lives in the past" (Moore, 2013). Thatcher's approach to "putting the great back into Great Britain," involved shifting it from a predominantly socialist welfare state to a society based on the individual and the free-market and was based in her assertion that there was "no such thing as society,"[27] only individuals, and that people must look to themselves first (and not the government). This approach involved drastic cuts in public spending that affected all areas from housing to local government, the privatization of formerly nationalized industries and sectors and the dismantling and (global) outsourcing of others (Hadley and Ho, 2010). The most infamous example involved the closing of "inefficient" mines in parts of Northern England, Scotland and Wales. Thatcher handled the consequent strikes, and the labor unions organizing them, with a militant contempt. Her calculation (not unlike many mainstream neoclassical economic thinkers) was that unemployment was an unavoidable fact of economic reform and that certain jobs (and workers) "would have to be the mulch that went into the revival of the general economic habitat" (Wood, 2019). It is widely believed that she tried out her most unpopular policies in Scotland first (e.g. the Poll Tax), though journalist James Wood (2019) suggests her treatment of Scotland (and Wales) had rather more to do with her possession of "what one colleague called 'a very English Englishness': she didn't sacrifice Scotland and Wales as part of a Conservative strategy; she hardly noticed they were there."

Thatcher's reforms left an indelible mark on all of the UK: "apart from the profound human misery that resulted, there was an enduring political cost—much of Scotland, Wales, and the North of England remains lost to Conservatives" (Wood, 2019). Cultural theorist and activist Stuart Hall argues that Thatcherism should be assessed not in electoral or economic terms but rather in terms of its role in reworking the "social imaginary" and "in disorganizing the labour movement and progressive forces, in shifting the terms of political debate, in reorganizing the political terrain and in changing the balance of political forces in favour of capital and the right" (1983, p. 13). As with the Clearances, it also re-ignited debates and conversation around the seemingly intractable issue of Scottish independence. Famous rock duo, The Proclaimers' (1988) song "Cap in Hand" addresses the widespread frustrations felt by many Scots at the time over the injustice of British rule, especially under Thatcher, and perhaps more potently, the submissiveness of their own people in the pursuit of self-rule.

The eventual successful vote for a devolved parliament in 1997 was the result of 15 years of cautious maneuvering under the patronage of the non-partisan "Campaign for a Scottish Assembly" and gave Scotland autonomy in "devolved" matters (e.g. education, health, environment and social services) as well as seats in the British Parliament. A referendum on Scottish independence, the Scottish National Party's *raison d'être*, finally came to fruition in 2014. In response to the single question, "Should Scotland be an independent country?" 55% voted "no" and 45% "yes." Many[28] believe that it is only a matter of time before Scotland becomes a sovereign nation. There is also no denying the impact that the (2016) "Brexit" vote—resulting in the UK voting to leave the EU—has had for the case for Scottish independence: Scots voted overwhelmingly against leaving the EU (62%) and many believe that this vote is another demonstration of the ways in which Scottish concerns are not represented in the British Parliament.

Decolonizing Scotland

The post-war period of decolonization and globalization saw the rise of new approaches to redressing and representing minority cultures. In Scotland, this saw a growing interest in the culture, language and traditions of the Gaelic speaking Highlands and Islands. This was associated with the founding of the School of Scottish Studies (now Celtic and Scottish Studies)[29] at the University of Edinburgh and the associated folk revival.[30] Co-founding member,[31] Hamish Henderson (of Perthshire), is credited with instigating the twentieth-century folk renaissance in Scotland. His notion of the "carrying stream of tradition" encapsulates emergent and radical approaches to "tradition" as a resource that, as Gary West puts it, "flows through time, picking up new flotsam as it goes, leaving some things on its banks in process" (2014, p.12–13) and folk culture, not as a quaint relic of the past, but as a living "manifestation of a rebel underground" (McFayden, 2019).

One of the most significant accomplishments of the School was to record and archive the oral histories and traditions of the Gaelic peoples of the Highlands and Islands before they disappeared. Because the Gaels were still a predominantly oral society during the time of the Clearances, official records of their experience were limited in large part to song and poetry, which, according to Auer (2015) emphasized the despair and shock associated with the loss of ancestral lands and, with them, the loss of their ties to ancestors. Poet and educator, Sorley Maclean (brother of the School's co-founder Callum Maclean), is a central figure in translating and immortalizing the life and struggles of Gaelic-speaking Scotland (though nearly 100 years after the last villages were cleared). Maclean was born on the Isle of Raasay[32] in 1911 and studied at the University of Edinburgh. In addition to a career dedicated to promoting and preserving Gaelic instruction in schools, Maclean is well known for his Gaelic poetry. His poem "Haillag" (1954)—named after an abandoned town on the Isle of Raasay that was cleared 100 years earlier—gave voice both to the sense of loss associated with the Clearances and to the past as it lingers in memory, poetry and place and refuses to be "cleared" or forgotten.

In addition to collecting, preserving and sharing the predominantly oral and intangible artifacts of Scotland's folk culture, Henderson and his colleagues at the School brought traditional singers and musicians from Scots and Gaelic speaking Scotland together in ceilidhs that were "a radical challenge to the Bing's Edinburgh Festival" and the elite twentieth-century idea that "the masses could be civilised by giving them access to culture that wasn't their own" (McFayden, 2019). According to McFayden, Henderson "championed the 'lived moment'," believing that poiesis (the poetic act of constantly "making the world new") occurs when people gather together in the spirit of conviviality. For Henderson, these were "the moments of 'resolve, transformation and insurrection', the 'proving ground for emotional and political truths' where the impulse and catalyst for resistance and change are to be found" (McFayden, 2019).

The carrying stream: conjuring and commoning Scotland's radical traditions

> The past is a foreign country: They do things differently there.
>
> (Hartley, 1953)

The oft-cited opening line of British novelist L.P. Hartley's (1953) novel *The Go-Between*, draws attention to complexities with which we experience, represent and use the past in the modern era and the problems inherent in our attempts to cement them in memory and history. As Hoare (2013) writes, "distant, intangible, unreliable, lost, our histories, at the levels of personal and national, are at best half-remembered and at worst actively misrepresented." Even so, the past is the re/source of the present and how we evoke it in the present remains a crucial political and sociological project. While the past can be a vital resource, it is also true that "too much nostalgia distracts us from the vital work—the darg—that we must do in the here and now, with an eye to the future" (McFayden, 2019). As Haiven and Khasnabish (2014) note, approaches to the "radical imagination" are based in an understanding of the past as a "commons" and history and memory as a process of "commoning." We believe projects like Tombreck Farm are symbolic of contemporary, radical approaches to "revival" that conjure the past not for its heroes, epics and distinctiveness but for its "commons."

As a project that seeks to regenerate a Highland community around the traditions of sustainability, conviviality and collaborative self-employment, Tombreck is evocative of contemporary commoning projects that extend Hamish Henderson's radical pragmatism—"We must start here, where we stand," he contended, "we can do no other" (in Neat, 2007)—to the creation of communities and economies *in place*. As economic geographer Gerda

Roelvink (2018) reminds us, when we frame our economies as a single capitalist system, "the struggle for justice is also placed on a global revolutionary scale, tied to the spatial and temporal imaginary of capitalism" (p. 133). Place-based approaches, on the other hand, move us from the revolution of a system to "connection through signification and a politics of assemblage" (ibid., p. 133). The idea of "assemblage" is rooted in the notion that "the commoning-community is more-than-human. The agent of change, the commoner, is no longer (and perhaps never was) a person or a category such as the working class but an assemblage" (Gibson-Graham, Cameron and Healey, 2013, p. 207). To think in terms of the assemblage is to be aware of, account for, and connect all the productive and reproductive resources that make our households, communities and nations work; including the more-than-human participants like socio-technologies, practices, work, habitats and other species (Roelvink, 2018).

One of the most striking features of Tombreck both as it presents itself, and as it operates, is its attention to the "assemblage" that makes it work—from the woodlands, to the people, to the principles and practices of permaculture and agroforestry, to the soil and earthworms, to the pigs and sheep, to the businesses and buildings—there is a keen awareness of the interconnections and interdependencies that constitute a rural farming ecological *and* economic system. As mentioned earlier in this chapter, the woodlands have always been an integral, indeed foundational, part of human habitation in the area. They remain so for Tombreck. As such, woodland regeneration was a founding component of creating Tombreck. Brown approached woodland generation using the principles of agroforestry, which are in many ways consistent with traditional approaches to woodlands in the region. In addition to mass planting indigenous trees (80,000 trees approximately covering a third of the farm), he fenced in some of the woodlands and grasslands in order to protect saplings from sheep and deer (Tombreck, 2017). In return, the woodlands provide building materials, natural boundaries between crops as well as a habitat for the farm's KuneKune pigs to forage and nest.

Building on the foundational commons of its woodlands, Tombreck is constituted by several site and land-based businesses and approaches to living and working that draw on the region's unique landscape and traditions.

"From cows to coriander"[33]

As founding community member Sue Manning indicated in an interview, the farm is the "backbone of it all"—for Tombreck to survive into the future, the farm needs to turn a profit (2022). That profit would in turn provide income and be put back into the "farm kitty" for use in upgrades and repair, enabling the community to weather changes and setbacks. While Tombreck currently has a small market garden and micro green business, Manning and Brown are seeking applications for a market garden and grazing tenancy. The vision for the market garden is that it produce both staples (e.g. potatoes and root vegetables) and "higher value leafy crops" for nine months to year-round production. The Group is open to various business models—including for example a cooperative, a community supported agriculture scheme or a box scheme—so long as it is based on mutuality and cooperation and an openness to involving the community. They also note the requirement that it implement the four International Federation of the Organic Movements principles of Care, Health, Ecology and Fairness.[34]

In order to complete their vision for a "small farm future," Manning and Brown are also seeking applications for a grazing enterprise on the farm that will "foster biodiversity, improve the health of the soils, improve productivity and become financially viable" (Tombreck, 2017). Their plan recognizes that the balance of grazing and fostering biodiversity "can be seen as a sliding scale with one subtracting from the other" and accordingly propose a "low intensity, grass based, native cattle grazing system is seen as being the most sustainable and suitable system to fulfil this vision" (Tombreck, 2017). Their vision calls for livestock being "outwintered" so that the farm's grasslands can have "long rest periods and short grazing periods" (Tombreck, 2017). They imagine cultivating small amounts of hay for on farm usage and the eventual inclusion of livestock into the native woodlands that would offer shelter and food and enable animals to exhibit natural behavior (Tombreck, 2017).

"It grew in our imaginations first"

The above quotation emerged in an interview with community members Sue Manning, Katy Macleod and Wendy Graham (2022) about the organic development of the "Big Shed" from a literal shed to a community center and centerpiece. Initially designed as a working shed for the farm (hence the name), it was later decided that a community building would better suit the needs of the local and wider community. The landholder, Tober Brown, gifted the land for the building to the Community and it is owned and managed through the The Lochtayside Community Interest Company (LTCIC), a nonprofit set up and run by Sue Manning for the Big Shed. Grant funding was awarded from the Big Lottery Fund, the Climate Challenge Fund, Perth & Kinross Council, SUST and Renewable Energy Scotland, for all of the capital costs.

The Big Shed was designed by Manning to be a model of "eco-design" with low carbon emissions and a reduced eco-footprint, heated from biomass and solar energy (Tombreck, 2017). The timber, mostly Norway spruce and Scottish larch, was locally sourced and the insulation is mostly sheep wool and hempcrete blocks. The walls are constructed of clay or lime plaster. The construction labor incorporated a Local Skills Training Programme, with local contractors, trades people and volunteers learning about eco-building techniques. The Big Shed opened to the community in September 2011, and it was awarded the first prize in the Carbon Trust's Low Carbon Building awards in the new build category in 2013 and was a runner up for another eco-building award (Tombreck, 2017). The building is managed by a small

Figure 4.2 The Big Shed at night.

group of volunteers, including interviewee Wendy Graham, who expressed her amazement at how far the idea had come in so few years.

The Big Shed is the literal and metaphoric centerpiece of the community: with a multipurpose hall, commercial kitchen, studio and workshop, its events and hires bridge the productive and reproductive needs of the community as they provide the means for income and community building and resilience. Over the past decade, the Big Shed has developed an established community with some groups and individuals returning regularly (e.g. weekly yoga classes, local scout group camps, kitchen hire, use of washing facilities and hire of the workshop to a local group). The Big Shed hosts concerts and musical events several times a year as well as yoga retreats and private hires (e.g. weddings, private parties, conferences, training events and studio hire) form the bulk of the income. In addition to being an important economic feature of the community, there is no understating the cultural and historical significance of the Big Shed as a place where the community can, as Hamish Henderson put it, engage in the poetic act of constantly "making the world a new" in lived moments of conviviality.

Our 2018 class visit of Tombreck was evocative of the pedagogical uses to which the site and the Big Shed are being put as a center for learning about local/rural sustainable development. Following a tour of the grounds, community members hosted us for discussion and tea and biscuits in the Big Shed in a moment of thoughtful, casual and convivial exchange. There is no underestimating the power that visits like these have on students. Indeed former students often comment on how much these moments of exchange and hospitality meant to them.

Collaborative self-employment and the spirit of communal self-reliance

One of the formative components of the 2003 Tombreck Action Group was an "employment/development proposal" that listed a series of " 'jobs' to be created in the spirit of 'collaborative self-employment'." The way in which these jobs were framed and described evoked a sort of "modern crofting" as interviewee and community member, Katy Macleod (2022) put it, including job sharing, cooperative enterprises and self-employment. When I asked Macleod what was meant by the term "collaborative self-employment," she replied,

> maybe just about us all working together sometimes, or some of us working together sometimes, for example, there have been occasions when the Big Shed has had a booking for a wedding and we've also been able to offer catering, photography and ceilidh music from amongst the residents.
>
> (2022)

Manning, Graham and Macleod described a recent instance involving the need for road works improvement on the property that also evokes the communal and adaptive approach to productive and reproductive work in the community. After sustained discussion, the community decided that they had the "skills, time and wherewithal" (Macleod, 2022) to do it themselves and that they did not want to be financially reimbursed for the work (Manning indicated her surprise at how strongly community members felt about this).

While the original list of jobs has evolved as the community has grown, the spirit of collaborative self-employment remains. The forms of labor (both paid and unpaid, productive and reproductive) performed, and types of businesses being run from the property, evoke the spirit of communal self-reliance mentioned by the interviewees; they also connect with the region's cultural and ecological past. Members work in local museums, craft shops and in local administration (post office, forestry, ranger), they run enterprises online and onsite that include willow growing, Scottish native wildflower growing, vegetable seed saving and production, an egg farm business, a small market garden, a micro green business and a beekeeping and honey business. Brown and Manning keep a small flock of Castlemilk Moorit sheep and a herd of free-range KuneKune pigs that graze within the agroforestry areas. Produce from the farm is sold through the "honesty box" at the Tombreck Farm Shop or local outlets and, as discussed earlier, all members take part in the fundamental reproductive labor required to maintain a sustainable and regenerative community.

Re-forming human nature

Scottish poet Norman MacCaig is perhaps best known for his Scottish nature poetry; however, the final stanza of his poem "Hotel Room, 12th Floor" (1966)—"the frontier is never somewhere else"—seems particularly relevant as it complicates the perennial habit of locating the frontier "somewhere else," somewhere other than our own nations and natures. It also problematizes the growing distance and detachment created by our economic and technological systems. While MacCaig's poem is bleak about the "midnight" that was, is, and will be human nature, there is something hopeful about knowing the cause of our suffering, that it is human nature that must be confronted and re-formed that human nature is nothing more than human subjects, in place and in relation and that reform might be as simple as resituating ourselves in the commonplaces—the "farm within farm" as MacCaig's "Summer Farm" puts it, that constitute us.

This is a task that is being taken on in various forms and formats all over the world. At Tombreck as elsewhere, people are "plotting and planting" new economic realities through a micropolitics of self-transformation and a macropolitics of the assemblage (Gibson-Graham, 2006b, p. xvi).

Notes

1 Plotting and planting is borrowed from Caroline Fraser's (2010) article about restoring Scotland's wild lands.

2 We would like to credit and thank Iain Stewart, tour guide extraordinaire turned friend, for introducing us to the community at Tombreck and for inspiring our interest in contemporary and historic Scotland, and colleague and friend Dr. Alexandra Peat for her personal insights and literary contributions to this chapter. We dedicate this chapter to you both (and take credit for all errors).

3 A "loch" is a lake and a "ben" is a mountain in Scottish Gaelic.

4 The quote is from the preface to *The Fair Maid of Perth* published in 1828.

5 Carie/Cragganester SSSI Wood are one of the largest areas of ancient, semi-natural wood in the area.

6 www.tombreck.co.uk/blank-2

7 Two homes, the Mill House and the North Steading, were sold.

8 See the following articles for five additional examples (www.scotsman.com/whats-on/arts-and-entertainment/5-alternative-communities-scotland-855090)

9 This quote comes from historical records of Highlander's who claimed their rights to the land being taken in the Clearances was based on ancestral relations to it (see Auer, 2015).

10 Crofting was/is a land tenure system of small-scale farming unique to the Scottish Highlands and Islands that emerged following the changes brought on by the Highland Clearances of 1750–1860 (see the Scottish Crofting Commission's: www.crofting.scotland.gov.uk/What-is-Crofting).

11 The buildings are owned by Tombreck Farm LLP.

12 There are discussions about expanding membership in the future.

13 Only Tober, Sue and WWOOFers contribute to the farm business. See https://wwoof.net/ for a description of the program.

14 Permaculture (a portmanteau of "permanent agriculture" and "permanent culture") is a systems-based design approach to agricultural ecosystems developed in the 1970s by Australian environmental researchers and educators Bill Mollison and David Holmgren that is intended as a corrective to industrial–agricultural systems and methods and is based in working with rather than against nature (see www.permaculturenews.org/).

15 This phrase emerged collectively in a group interview and no one remembers who said it first.

16 The community is still researching different models and Manning indicated that neither the Farm nor the community are ready.

17 Colleague Alexandra Peat importantly reminded me that hikers still die in the Scottish hills every year … it's not that tame.

18 See https://treesforlife.org.uk/

19 See https://reforestingscotland.org/what-we-do/the-reforesting-scotland-vision/

20 See The Highland Council, n.d.

21 See Simpkins (2019) for a contemporary take on the geographical divides in Scotland as they influence views on independence.

22 The Highland Boundary Fault is a major fault zone that separates two geological terranes and has given way to two geographical terrains that are characterized by a visible change in topography.

23 They were also a battle over monarchal issues of succession, the notion of the "divine rule of kings," and differing political and religious outlooks.

24 Predominantly to the colonies of Australia, Canada, the United States and New Zealand.

25 For example, Burns famously went on a Highland Tour in 1787 that inspired some of his most popular poems and songs including the song, "The Birks of Aberfeldy," which was penned during his stay in Aberfeldy, not far from Tombreck and Scott is famous for orchestrating King George IV's 1822 visit to Scotland in an event filled with Highland pageantry and regalia (and pink tights!) that is said to have instigated the tradition of "Balmorality"—a term coined by writer George Scott-Moncrieff in 1932 to describe the superficial romanticization and cultural appropriation of (Highland) Scottish identity and traditions associated with Queen Victoria and Prince Albert and their 1852 purchase of Balmoral Castle in Highland Scotland.

26 A Scottish "Home Rule" bill was first presented to the British Parliament in 1913 but was stalled due to the emerging war efforts. Whereas Ireland fought and won its War of Independence in 1922, the Scottish took a different tack creating the non-partisan Scottish Covenant, who drafted a petition first proposed in 1930, written in 1949 and was eventually signed by more than 2 million Scots but which had little effect (BBC, 1999). The Scottish National Party was formed in the mid-1930s around the cause of independence and in reaction in part to the Labour Party's distancing from the cause of independence (BBC, 1999). Home rule did not enter the political mainstream again until the post-war processes of decolonization took hold.

27 These are quotations from an interview with Woman's Own. See Thatcher (1987)

28 Most prominently those associated with the Scottish National Party.

29 Henderson and Maclean were the first ethnographers brought on to collect and study folk culture at the university. Both had collaborated with American folklorist, Alan Lomax, who is credited as the catalyst and inspiration for the work of the School of Scottish Studies.
30 Which began with the "Edinburgh People's Festival Ceilidhs" in the 1950s.
31 With Calum Maclean, brother of poet Sorley Maclean.
32 The entire populations of 12 townships were cleared from isle of Raasay between 1852 and 1854.
33 Sue Manning coined this phrase in our June 2022 interview.
34 See www.tombreck.co.uk/_files/ugd/60e00d_0427e3e6be9c4df99d57a4fe19b0aecb.pdf

References

Adams, W. and Mulligan M. (2004) *Decolonizing nature: Strategies for conservation in a post-colonial era*. London: Earthscan Publications.

Auer, C. (2015) The representation of land in the Gaelic poetry of the Clearances. In Laplace, P. (ed.). *Environmental and ecological readings: Nature, human and posthuman dimensions in Scotland*. Besançon: Presses Universitaires de Franche-Comté, pp. 61–73.

Bevan-Jones, B. (2004). *The ancient yew*. Macclesfield: Windgather Press.

Burns, R. (1785) To a mouse. *Scottish Poetry Library*. Available at: www.scottishpoetrylibrary.org.uk/poem/mouse/

Campbell, A. (2000). *A history of the Clan Campbell*. Edinburgh: Edinburgh University Press.

Chambers, R. (1869) *History of the Rebellion of 1745–6*. London and Edinburgh: W. & R. Chambers.

Devine, T.M. (1983). Highland migration to lowland Scotland, 1760–1860. *The Scottish Historical Review*, 62(174), pp. 137–149.

Devine, T.M. (2006) In bed with an elephant: Almost three hundred years of the Anglo-Scottish union. *Scottish Affairs*. Institute of Governance, University of Edinburgh (57), pp. 1–18.

Devine T.M. (2018) *The Scottish Clearances: A history of the dispossessed 1600 to 1900*. London: Allen Lane.

Dress Act (1746) Great Britain. In Bentham, J. (ed.). *The statutes at large, from Magna Charta to … 1869 …* Cambridge: Oxford University. Available at: https://archive.org/details/statutesatlarge01britgoog/page/526/mode/2up

Fraser, C. (2010) In Scotland's search for roots, a push to restore wild lands. *YaleEnvironment360*, Yale School of the Environment. Available from: https://e360.yale.edu/features/in_scotlands_search_for_roots_a_push_to_restore_wild_lands

Gibson-Graham, J.K. (2006) *The end of capitalism (as we knew it): A feminist critique of political economy*. Minneapolis: University of Minnesota Press.

Gibson-Graham, J.K., Cameron, J. and Healy, S. (2013) *Take back the economy: An ethical guide for transforming our communities* Minneapolis: University of Minnesota Press.

Graham, W. (2022, June 7). Interview.

Graham, W. (2022, June 21). Email correspondence.

Hadley, L and Ho, E. (2010). *Thatcher and after: Margaret Thatcher and her afterlife in contemporary culture*. London: Palgrave Macmillan.

Haiven, M. and Khasnabish, D.A. (2014) The radical imagination: Social movement research in the age of austerity. London: Bloomsbury Publishing.

Hartley, L.P. (1953). *The go-between*. London: Penguin (originally Hamish Hamilton).

Hechter, M. (1975) The modern world-system: Capitalist agriculture and the origins of the European world-economy in the sixteenth century. *Contemporary Sociology*, 4(3), pp. 217–222.

Hechter, M. (2021) Internal colonialism, alien rule, and famine in Ireland and Ukraine. *East/West: Journal of Ukrainian Studies*, 8(1), pp. 145–157.

The Highland Council. (n.d.) Regeneration. Retrieved on June 22, 2022 from www.highland.gov.uk/info/20014/economic_development/844/regeneration

Hoare, N. (2013) The past is a foreign country. Retrieved on June 23, 2022 from www.thewhitereview.org/feature/the-past-is-a-foreign-country/#:~:text='THE%20PAST%20IS%20A%20FOREIGN,inherent%20to%20memory%20and%20history.

Laplace, P. (2015). *Environmental and ecological readings: Nature, human and posthuman dimensions in Scotland*. Besançon: Presses Universitaires de Franche-Comté.

Manning, S. (2022, June 7). Interview.

McCaig, N. (1993) Hotel room, 12th floor. *Norman MacCaig: Collected poems*. London: Chatto & Windus. Retrieved from www.scottishpoetrylibrary.org.uk/poem/hotel-room-12th-floor/

Macleod, K. (2022, June 7) Interview.

Macleod, K. (2022, June 6) Email exchange.

McFayden, M. (2019, August 16) Hamish Henderson and the liberated life. *Bella Caledonia*. Retrieved on June 15, 2022 from https://bellacaledonia.org.uk/2019/08/16/hamish-henderson-and-the-liberated-life

McQueen, J. (1974) The work of the School of Scottish Studies. *Oral History* (2)1, pp. 62–64. Retrieved on June 15, 2022 from www.jstor.org/stable/40178400

Mitchell, J. (2014) *The Scottish question*. Oxford: Oxford University Press.

Moore, C. (2013) *Margaret Thatcher: The authorized biography, volume one*. London: Penguin.

Neat, T. (2007). *Hamish Henderson:Aa biography. Vol 1: The making of the poet*. Edinburgh: Polygon.

Press Team Scotland (July 17, 2013). "Growing a community asset." Big Blog Scotland. Retrieved from: https://bigblogscotland.org.uk/2013/07/17/growing-a-community-asset/

Roelvink, G. (2018) Community economics and climate justice. In Jacobsen, S.G. (ed.). *Climate justice and the economy: Social mobilization, knowledge and the political*. London: Routledge, pp. 129–147.

Scott, W. (1871) *The Fair Maid of Perth, or, St. Valentine's Day: chronicles of the Canongate second series*. Edinburgh: Adam and Charles Black.

Simpkins, F. (2019) Twenty years of devolution in Scotland: The end of a British party system? *Review Français de Civilisation Britannique, XXIV-4*, https://doi.org/10.4000/rfcb.4938

Solly, M. (2020) A not-so-brief history of Scottish independence. *Smithsonian Magazine*, January 30.

Thatcher, M. (1987, September 23). Interview with Woman's Own. Retrieved on June 18 from www.margaretthatcher.org/document/106689

Todaro, L. (2022, April 22) Scotland might become the world's first rewilding nation. *AFAR*. Retrieved on May 20 from www.afar.com/magazine/scotland-might-become-the-worlds-first-rewilding-nation

Tombreck. (2017) Tombreck. Retrieved on May 1, 2022 from www.tombreck.co.uk/

West, G. (2014) *Voicing Scotland: Folk, culture, nation*. Edinburgh: Luath Press Limited.

Wood, J. (2019, December 2) It's still Margaret Thatcher's Britain. *The New Yorker*.

Wright, S., Cairns, P. and Underdown, N. (2018). *Scotland a Rewilding Journey*. Kingussie, Scotland: Scotland–The Big Picture.

5 Seed change

Navdanya and the reimagination of reproductive economies

Introduction

Navdanya, a movement started by Vandana Shiva, has become one of the most well-known movements in the areas of agroecology and bio-conservation. Shiva, who is a trained physicist, was greatly inspired by the Chipko movement in the 1970s. The Chipko movement originated in the Garhwal region of India and is situated in the foothills of the Himalayas. It stems from the "Forest Satyagraha" movements that emerged in colonial India in response to the enclosure of forests for commercial logging and agriculture. Prior to such enclosures, forests were managed as common resources "with strict informal, social mechanisms for controlling their exploitation to ensure sustained productivity" (Shiva and Bandyopadhyay, 1986, p. 134). The most famous actions under this movement took place in the forests of Adwani, Amarsar and Badigarh in the Garhwal region from 1977–1978, when local women hugged the trees[1] of the forest to protect them from contractors sent to cut them for commercial use.

Shiva started Navdanaya in 1987. The organization's webpage describes Navdanya as an "earth centric, women centric, farmer led movement for the protection of biological and cultural diversity" (Navdanya, n.d.). Inspired by the Gandhian philosophy of non-violence, Shiva began Navdanya in the state of Karnataka and region of Tehri Garhwal in the state of Uttarakhand as a seed saving movement intended to preserve seeds and indigenous knowledge. The word Navdanya means "nine seeds" and is meant to symbolize both cultural and biological diversity. These "nine seeds"—barley, millet, pigeon pea, mung, chickpea, rice, sesame, black gram and horse gram—form the basis of self-sufficiency.

Navdanya was registered as a trust in 1991. In 1994, the first community seed bank was started in Garhwal and quickly thereafter, the Navdanya Biodiversity farm was formed. In 2001, the Earth University (Bija Vidyapeet) was launched. The campus serves as a center for providing information and knowledge exchange with, farmers and consumers about organic farming. It also serves as an experimental farm to study and understand different methods of organic farming. Today the campus boasts a water harvesting

DOI: 10.4324/9781003138617-5

facility for irrigation, compost shed for soil fertilization, a library, a lecture hall and a laboratory for soil testing. The campus also hosts students and practitioners from all over the world who are interested in learning more about seed keeping as well as techniques for organic farming. The university offers regular courses for interested students on the subjects and practices of sustainable farming and biodiversity. Students can also volunteer at the site and learn more about sustainable agriculture. The campus grows its own food and uses biogas to cook food. Volunteers and students are expected to clean the campus, wash the dishes and tend to the crops as a part of their study. Navdanya International, located in Rome, Italy, was established in 2011 to expand the reach of the movement. The organization started the Global Movement for Seed Freedom to bring together different organizations around the world involved in saving seed and promoting ecological agriculture.

Navdanya emerged in response to the ecological damage caused by the so-called "Green Revolution" in India when specific (high-yield) varieties of seeds were promoted. High yield variety (HYV) seeds were first introduced in India following a severe drought in 1966 via the Fourth Five-Year Plan under the leadership of agricultural scientist M.S. Swaminathan. The drought led to severe food scarcity and forced India to import grains from the USA. Various measures were undertaken to revitalize the Indian agricultural sector. There was special emphasis on the use of fertilizers, electricity and diesel oil to increase production of food grains. In addition, attention was paid to minor irrigation facilities and to a more intensive use of HYV seeds. According to Chakravarty (1987), the introduction of the Fourth Five-Year Plan led to some key changes in the agricultural sector in India. These changes increased the use of purchased input, greater monetization of Indian agriculture, introduction of price support policies and greater use of oil-based fertilizers. These transformations also saw intensified agriculture-industry linkages and an

Figure 5.1 Earth University.

increased sensitivity to international fluctuations in world prices of oil. The "Green Revolution" also resulted in a decrease of biodiversity[2] as only specific varieties of rice and wheat were promoted by the government.

Navdanya is a direct response to these changes that seeks to save indigenous seeds and help farmers break out of farming based on high input costs and increased vulnerability to fluctuations in international markets. The organization has also actively protested the introduction and use of Genetically Modified Seeds in the country. The organization's mission includes empowering marginalized communities by making them self-reliant through the sustainable and fair use of natural resources, promoting peace and harmony through the conservation, renewal and rejuvenation of the "gifts of biodiversity" and defending these gifts as commons, and creating living economies through living democracies. The movement is also committed to: seed sovereignty (Bija Swaraj) through seed banks, food sovereignty (Ann Swaraj) through the education of farmers about the ill-effects of using chemicals in agriculture, living soil (Bhu Swaraj) through organic and agroecological approaches to farming, knowledge democracy (Gyan Swaraj) the promotion of scientific research on biodiversity, agro-ecology and finally through challenging attempts to patent traditional and indigenous knowledge (Navdanya, n.d.).

Seeds and reproduction

> Without seed there is no food. Without food there is no life. Saving seeds is saving life in all its diversity.
>
> (Seed Keeper, Navdanya)

As mentioned above, Navdanya operates through community seed banks. According to Shiva (2014), "seed is not the just the source of life. It is the very foundation of our being" (n.p.). She further explains that for thousands of years, farmers (particularly women farmers) have been involved in breeding and evolving seeds to increase their diversity. Women have worked together for generations in conjunction with nature to protect and preserve seeds to meet the demands of their communities. Shiva asserts, "biodiversity and cultural diversity have mutually shaped one another" (n.p.).

In a report produced by Navdanya (2012), Shiva argues that there has been an "assault on seed" (p. 2) as evidenced by the rapid decrease in seed diversity. Much of this erosion has been due to concentration of control of seeds by a few corporations and monoculture. This already worrying situation was made worse with the Trade Related Intellectual Property Rights (TRIPS) agreement in 1995. Article 27.3 (b) of the TRIPS agreement states that members may also exclude from patentability:

> plants and animals other than micro-organisms, and essentially biological processes for the production of plants or animals other than non-biological and micro-biological processes. However, Members

(capitalization intended) shall provide for the protection of plant varieties either by patents or by an effective sui generis system or by any combination thereof.

(WTO, n.d)

The debate around this article has been intense. While special interest lobbies in developed countries support this article, as it gives the opportunity to multinationals to patent life forms including seeds,[3] developing countries like India, Brazil and many African nations have protested the article understanding it endangers, and in some cases even criminalizes, farmers' right to traditional knowledge.

Navdanya describe patents on life and new biotechnologies as today's tools of imperialisms. In doing so, they have challenged some of the fundamental values of international trade and the WTO's rules regarding TRIPS. They assert: "A patent is an exclusive right to own, make, sell, produce, use a patented product. A patent on seed implies that a farmer saving seed is an 'intellectual property thief.' But it means more. A system in which seed has become a corporate monopoly, a system in which a few companies control the seed supply is in effect a system of slavery for farmers. Where the freedom of seed disappears, the freedom of farmers disappears" (Navdanya, 2016). Shiva (2012) further argues that,

> a reductionist, mechanistic science and a legal framework for privatizing seed and knowledge of the seed reinforce each other to destroy diversity, deny farmers innovation and breeding, enclose the biological and intellectual commons, create seed monopolies. Farmers varieties have been called land races, primitive cultivars. They have been reduced to a "genetic mine" to be stolen, extracted and patented. Not only is the negation of farmers' breeding unfair and unjust to farmers, it is unfair and unjust to society as a whole.
>
> (p. 2)

The patenting and commercialization of seeds has far reaching implications for small farmers as it increases input costs for them. According to The Tricontinental (2019), over 300,000 farmers have committed suicide in India since 1995 due to high levels of indebtedness. According to Shiva (2014), farmers in India have become increasingly dependent on nonrenewable seeds. She gives the example of Monsanto's increasing hold on the supply of cotton seeds in India. Control of seeds by a company like Monsanto has led to significant shifts in the supply of seeds. The company, through its various joint ventures and licensing arrangements, now controls a significant portion of cotton seed markets. According to Shiva (2014), this near monopolization and commodification of seeds has converted them "from a renewable regenerative, multiplicative resource into a nonrenewable resource and commodity" (n.p.). She further explains that this conversion of seeds by Monsanto to intellectual property may provide the company a claim to the profits made by farmers through royalty payments, thereby increasing the levels of debt for farmers. It is noteworthy that, according to a report by Center for Human Rights and Global Justice (2011), farmer suicide has been disproportionately high in the cotton sector.[4] Yet another impact of the adoption of commercial seeds by farmers has been the loss in biodiversity, as cotton is now cultivated as a monoculture crop.

Navdanya's flagship program, "Community Seed Banks," offers farmers the opportunity to save different varieties of indigenous seeds in common banks. The organization started its first seed bank in their biodiversity farm in the village of Ramgarh near the city of Dehradun in the state of Uttarakhand. When the land was acquired in 1996, it was infertile and dry due to excessive cultivation of sugar cane and eucalyptus.[5] The first seed bank was started by Shiva to collect and store indigenous varieties of seeds that were collected from farmers to ensure that these seeds do not disappear from Indian agricultural communities. The idea of seed banks for Shiva is not to, as she says, "museumize" seeds for future generations but rather re-introduce them in agricultural cycles.

As of 2020, Navdanya has helped start 144 community seed banks in 22 states in India (Seed Freedom, 2020). The organization was also involved in the identification of local seed keepers in different agro-climactic zones of India. Navdanya works with local farmers through these community seed banks to preserve indigenous varieties of seeds and to preserve local knowledge about indigenous farming techniques that tend to be chemical free in nature.

Figure 5.2 Seed varieties showcased at Earth University for educational purposes.

Figure 5.3 Record of increases in seed varieties from 2010–2018.

These community seed banks are managed by the community members themselves, who offer local farmers seeds that are indigenous to the agro-climatic areas and therefore more resilient to the climate of the region. The seed banks act as real banks that loan out traditional indigenous seeds to interested farmers. The farmer is then expected to return the seeds to the seed bank with an additional number of seeds (around 25%). They also have the option of giving their seeds for free to two other farmers who are expected to do the same in the next cycle of harvesting. Many community seed banks now operate independently after the organization's initial support. Navdanya has trained close to 750,000 farmers so far (Navdanya, 2016). These community seed banks have saved up to 4,000 rice varieties and many indigenous varieties of millets and pulses.

These community seed banks have helped rebuild distressed communities after major natural calamities. For example, in the state of Odisha on the eastern coast of India, the community seed bank provided farmers with salt-resistant rice after the area was hit by a super cyclone in 1999. The seed bank also distributed salt-resistant paddy seeds to other areas affected by the super cyclone including to communities in Tamil Nadu and West Bengal (Shiva, 2016). Another example involves the cotton growing region of Vidarbha in the state of Maharashtra, where farmer indebtedness due to use of commercial seeds and pesticides has led to a spate of farmer suicides (Center for Human Rights and Global Justice, 2011). Navdanya has partnered with the Maharashtra government to provide indigenous cotton seeds to farmers and to help them grow crops that are more resilient to the local climate (Navdanya, 2016).

The seed keepers who manage community seed banks are from the local communities and many are women. Navdanya believes that women do the bulk of the activities in agriculture starting from sowing, harvesting and maintaining kitchen gardens, to maintaining and preserving seeds and indigenous knowledge. These seed keepers are keenly interested in preserving and expanding their knowledge of local and indigenous seeds. In a conversation with a seed keeper expert at the biodiversity farm, one seed keeper said: "I am not interested in food, but I always ask for seeds from anyone I meet."[6] This seed keeper had approached Shiva when she started Navdanya looking for work; however, when Shiva found out that she was a farmer, she was asked to help the organization with seed preservation (Seed Freedom, n.d.). She has been instrumental in preserving and disseminating knowledge regarding local and indigenous seeds from the very start of the organization.

Seed keepers are also trained by Navdanya in the areas of vermicompost, crop rotation and harvesting. These seed keepers then provide training to the farmers from their villages. In the region of Vidarbha, for example, farmers received training on how to save, select and produce seeds. They were also given training on vermicompost, soil examination and regenerative organic agriculture. The one common theme that emerged out of the interviews with these farmers was their enthusiasm to teach their fellow farmers and spread the knowledge they had received to their communities (Navdanya, 2020).

Navdanya has also been instrumental in raising awareness regarding the dangers of the commercialization of agriculture and seeds in India.

Colonization of agriculture: the agrarian crisis

The emergence of movements like Navdanya must be viewed in their historical context. For Navandya, the crisis of Indian agriculture caused by British imperialism is central. British colonialism fundamentally changed the structure of Indian agriculture. It introduced new forms of taxation and monetized the agricultural sector (Bagchi, 1982). The introduction of the zamindaari[7] system through the Permanent Settlement Act in 1793 gave the zamindaar class the right to collect revenues thereby giving them effective ownership of the land and dispossessing peasants from the land (Alavi, 1975). Further, commercialization of agriculture through the new monetized system of taxation forced farmers to grow cash crops for exports making them vulnerable to price fluctuations in the international markets. According to Alavi (1975), British rule also destroyed the self-sufficiency of Indian villages. Villages were integrated with the world economy through trade. In many cases, farmers were forced to produce cash crops like indigo, which destroyed the fertility of the soil. The Champaran protests in 1917 was the first Satyagraha led by Gandhi against the cruel practices used by landlords on farmers to force them to produce indigo. Alavi (1975) further argues that with British colonization, Indian agriculture generated surplus that was extracted for capital accumulation in the core (i.e. England).

In post independent India, the implementation of the Green Revolution in response to food shortages and to a nationalistic fervor for food sovereignty, led to a bumper crop in 1968. It was showcased as a great victory of scientific advancement over nature. However, there were critics who saw the Green Revolution for what it was. Referring to the Green Revolution, Harry Cleaver wrote in 1972,

> The development of this new technology is very much a part of the efforts of the American elite to direct the course of social and economic development in the Third World. With a foreign policy devoted to facilitating the expansion of U.S. multi-national business, the elite is always concerned with creating new investment and sales markets.
>
> (p. 177)

Since the end of the Second World War, the US was providing food aid to Third World nations on an ad-hoc basis during "socially disruptive food shortages" (Cleaver, 1972, p. 179). Food export was continued under P.L. 480 from 1954. However, President Lyndon Johnson reversed this policy in 1965 declaring that the US would not hand out their surplus grains for free anymore.

In the case of India, as mentioned earlier, concurrent droughts and food shortage forced Prime Minister Indira Gandhi to open markets to US capital. As Patnaik (2020), puts it:

> The country's import-dependence became so great that it was literally a case of "ships-to-mouth"; and arm-twisting by the United States, a clear

> example of what some have come to call "food imperialism", became so odious that Indira Gandhi asked Jagjivan Ram, the then Food and Agriculture Minister, to expedite the "Green Revolution.
>
> (n.p.)

There were concerns that the increased mechanization of agriculture would lead to displacement of labor. The main beneficiaries of this plan were big farmers who were able to access credit to invest in new technology. The impact of the Fourth Five-Year Plan on labor absorption remains indeterminate. For example, Osmani (1993) explains that macro-level evidence showed that per-hectare labor use in states like Punjab, Haryana and Uttar Pradesh—the three states where the impact of Green Revolution was maximum—had fallen during this time, but labor absorption had actually increased in the same period. This has been attributed to the increase in the gross cropped area that was direct result of better irrigation facilities at this time. According to Bhalla (1989), the labor absorptive capacity of Indian agriculture saw an expansion from the mid-sixties to the mid-seventies. This expansion was primarily due to the introduction of high yielding variety seeds, extension of irrigation facilities and use of fertilizers. She also points out that employment elasticities for some of the crops like wheat were actually negative during this time. She also observed that labor absorption in some of the best performing states had followed an "inverted-U" shaped path. According to her,

> The initial response to "Green Revolution" technology was a sustained rise in labour use per hectare. This trend characteristically peaked in the mid-seventies or shortly afterwards, and the subsequent increases in yield were associated with a contraction in man days employment per hectare, in case of most crops
>
> (Bhalla, 1989, p. 4)

Overall, the impact of the Green Revolution in India has been mired in debate. Some hailed the movement a life saver for India by increasing the productivity of land in the states of Punjab and Haryana and thereby making the country food independent. Others argue that the Green Revolution led to loss in soil fertility in these states in the recent times and devastated the local eco-system. For example, Shiva (2016) describes Punjab as a state that has been left with infertile soil, crops that are prone to insects, waterlogged deserts and farmers who are heavily indebted.

Initially, the Green Revolution was limited to the irrigated states of Punjab and Haryana but by the 1970s, it spread to semi-arid areas where farmers were often encouraged to cultivate cash crops like cotton that requires large amounts of water. The spread of tube wells and pump sets to extract water for such water intensive crops have led to a decline in ground water levels in these regions (Walker, 2008). The Green Revolution also led to an increase in the land under cultivation. As Nadkarni (1996) observed the initial high returns on agriculture led to greater encroachment of forests and privatization of common resources. In addition, the Green Revolution also resulted in changes in the class structures of rural India, where the rich farmers were able to take advantage of the government policies, while the poorer farmers became wage laborers. But these wage laborers were further exploited through extra-economic forces like the debt-bondage (Bhaduri, 1983).

The introduction of HYV seeds into Indian agriculture must be seen within the historical context of a post-war world that was divided into two main blocs as a result of the cold war. According to Ajl and Sharma (2022), the US's push for the introduction of modern technology in India was a result their deep fears regarding the "red revolution" in the post-colonial world. American organizations like the Rockerfeller and Ford Foundations, US Aid, and the World Bank were deeply worried about the revolutionary capacities of small peasant movements across Asia during the period of decolonization. As noted by Anderson and Morrison (1982) the peasantry were considered "incipient revolutionaries and if squeezed too hard could be rallied against the new bourgeois-dominated governments in Asia" (p. 2). Cleaver (1982) adds, "Food was clearly recognized as a political weapon in the efforts to thwart peasant revolution in many places in Asia" (p. 269). Shiva (1991) argues that while the Green Revolution was promoted as a way to increase food production and reduce agrarian conflicts, it resulted in "new relationships between science and agriculture, defined new links between the state and the cultivators, between international interests and local communities, and within the agrarian society" (p. 47). Thus, we see that the implementation of the Green Revolution in India was a direct result of American imperialism in its quest to increase its influence over post-colonial nations.

In the post-liberalization period,[8] the agrarian crisis deepened due to the neoliberal policies pursued by the government. The government reduced its public expenditure to meet the conditions set by agencies like the World Bank and the International Monetary Fund in return for international loans (Sen and Ghosh, 2017; Patnaik, 2004). During this period, agriculture was characterized by stagnation due to reduced government spending in the sector, discontinuation of social welfare programs and a move towards privatization. In 1998, the seed sector in India was further privatized with a greater push to replace saved seeds with corporate seeds, which were dependent on more intensive use of fertilizers and pesticides and more importantly could not be saved. The entry of genetically modified seeds sold by large corporations in the cotton growing regions of India has already been discussed above.

Further, the reforms in India also pushed for greater privatization of land for real estate development and the establishment of special economic zones (SEZs) thereby increasing the demand for non-agricultural use of land. This has led to many small and marginal farmers being disposed from their lands. This is what Harvey (2005) describes as "accumulation by dispossession," which is fundamental to capitalist expansion. According to him, accumulation by dispossession is different from the expansion of capitalism through expansion of wage labor as the latter produced a counter " oppositional culture" of trade unions and working class political parties. However, dispossession, "is fragmented and particular—a privatization here, an environmental degradation there, a financial crisis of indebtedness somewhere else" (p. 178), thereby making it difficult to oppose.

In more recent times, the government of India passed[9] the Farm Bills of 2020. These bills were supposedly aimed at liberalizing the agricultural sector by bringing farmers in direct contact with corporate buyers. The government's stated objective was to make the agricultural sector more competitive, thereby increasing the incomes of farmers. However, the bills were strongly opposed by many different farmer groups who argued that the bills would leave them exposed to the mercy of large corporations as the government quietly retreats from the agricultural sector. These farmers' protests eventually forced the government to repeal these laws.[10]

Yet another issue faced by the agricultural sector was a credit crisis, caused by the deregulation of the banking sector after liberalization. Prior to liberalization, agriculture and small-scale industry were treated as priority sectors for loans at low interest rates. But after liberalization, Indian banks were not required to prioritize these sectors, and were thus free to pursue profits in other more developed sectors (Patnaik, 2004). This created a crisis of accessing finance by small farmers, thereby pushing them to borrow money from local money lenders, who typically offer loans at very high rates of interest. This is one reason for the severe indebtedness of Indian farmers discussed above. Liberalization of the economy and trade liberalization also exposed farmers to the world commodity markets and incentivized them to grow cash crops. For example, as world cotton prices increased in the early 1990s, many farmers switched to cotton in hopes of high profits. Because many were inexperienced with the crop, the crash of commodity prices in mid-90s resulted in huge losses for the cotton farmers further increasing their indebtedness (Patnaik 2004)

When we examine the evolution of Indian agriculture since colonial times, we see that that there have always been attempts to colonize this sector most of which have been successful. While in the colonial period the British Empire forcefully commercialized this sector to extract surplus for its own capitalist production, in the post-independence period, the Indian government found itself in a vulnerable position at different points in time, which forced it to open the agricultural sector to foreign corporations. The colonization of agriculture by foreign entities has been a continuous process in Indian history.

Decolonizing and recommoning

> Our vision is based on the imperative to create alternatives to gigantism and centralisation by building on the strengths of the small and the decentralized and for sustainability and justice, for freedom and peace.

> A globalisation of the local, rather than a globalisation driven by giant corporations and institutions is our vision.
>
> (Navdanya, 2016)

In addition to its influences on agriculture, colonization also led to ecological, cultural and social crises. The Green Revolution, for example, not only resulted in a significant loss of biodiversity, it also erased with it centuries of collective knowledge. Referring to Agricultural Development Council President Arthur T. Mosher's[11] handbook on US agricultural policies, Cleaver (1972) points out the recurrent "theme of teaching peasants to want more for themselves, to abandon collective habits, and to get on with the 'business' of farming" (p. 179). Both the Green Revolution and the more recent neoliberal attempts to commercialize agriculture in India have resulted in the loss of the collective knowledge that farmers had, until these regimes, preserved and disseminated for generations.

Shiva (2014) calls the domination of " western scientific knowledge," the "Monoculture of the Mind." She further explains: "the disappearance of local knowledge through its interaction with the dominant Western knowledge takes place at many levels, through many steps. First, local knowledge is made to disappear by simply not seeing it, by negating its very existence" (n.p.). She further explains that,

> The knowledge and power nexus is inherent in the dominant system because, as a conceptual framework, it is associated with a set of values based on power that emerged with the rise of commercial capitalism. It generates inequalities and domination by the way such knowledge is generated and structured, the way it is legitimized and alternatives are delegitimized, and by the way in which such knowledge transforms nature and society. Power is also built into the perspective that views the dominant system not as a globalized local tradition but as a universal tradition, inherently superior to local systems.
>
> (n.p.)

The colonization of agriculture is, simply speaking, also the colonization of minds and the erasure of indigenous knowledge. The erasure of indigenous knowledge has led to the present ecological, social and cultural crisis. In other words, it has resulted in a crisis of reproduction. And this crisis is a direct result of commercialization and privatization of the agricultural sector. Finally, it has led to a crisis for farmers (labor) who have been trapped in a cycle of production for the markets but have no resources to reproduce their own selves and communities.

Movements like Navdanya[12] through their seed banks are trying to decolonize and common the fundamental means of production: the seed. By encouraging the preservation and spread of the indigenous knowledge, Navdanya is trying to decolonize and common social reproduction. Navdanya not only challenges but rejects the dominant capitalist system as a feasible form of economic organization, rejecting the hierarchies that capitalism creates between nature and humans and challenging the dominant language of economics that has reduced nature to a simple factor of production that can be exploited without limit. A letter from "Mother Earth" to her "children" on the Navdanya website directly addresses and challenges the ways in which the current economic paradigm conceives of the relationship between nature, humans and the economy, as in this excerpt:

> Dear children,
>
> I embrace you and welcome you back home .
>
> You were uprooted and thrown into the cities by an unfair, unjust, violent economy which made you think that taking care of me was "inferior work", work that humanity could stop doing. They hid the falsehoods of the economy of limitless greed, driven by the illusion of limitless economic growth, and made you blind to my limitless generosity and abundance as long as you receive my gifts within the ecological limits I have created through my laws.
>
> (Navdanya, 2016)

By re-centering nature symbolically (as in this letter) and literally (as in the seed banks) Navdanya rejects mainstream economic language and makes nature an author of the system rather than a mere backdrop (exogenous variable).

The letter continues:

> In violating my laws and my Rights, a small group among my children, also robbed you of the rights I have given you—the right to life, to food and water and air, the right to work , to be creative, to find meaning. My spoilt children created constructs like GDP and made them the "law" of the "Market" to hide their limitless greed justify their limitless violence against me, and you ,the rest of my children. They sold toxic Poisons which pushed most of my children to extinction. and extracted super profits from you by selling you unnecessary, costly non renewable seeds, and toxic chemicals. They told you to stop caring for me and protecting biodiversity to grow food that nourishes me, and your family, your community, and the country.
>
> (Navdanya, 2016)

As can be seen from the above excerpt, Navdayna challenges and rejects the rhetoric of capitalist dominance. The project also reimagines the language of economics. By recentering nature and focusing on its role in the economy, the organization is attempting to decolonize the language used in mainstream economics. Specifically, it highlights the inherent contradictions in the ways in which mainstream economic theory talks about progress and development but uses methods that are detrimental to nature and ecosystems, as if human development can occur in a vacuum and outside of nature.

The movement has made a deliberate attempt to distance itself from the dominant economic discourse by using language that decolonizes as well as engages in commoning all stakeholders of an economy that includes the farmers and nature.

In order to better understand how Navdanya provides us a template for reproductive economies by reimaging language and practices, we examine the notions of the seed keepers the seed banks and interest.

As mentioned above seed keepers, also called *bijaks*, are local farmers who manage the seed banks, share seeds with other farmers as well as knowledge about different farming techniques. The deliberate use of the term "keeper" is interesting. A keeper is not an owner. The word keeper typically involves the care of something or someone. Referring to farmers as seed keepers shifts the focus to care rather than private ownership. In other words, it challenges the language of ownership.

The following quotes from seed keepers illustrates this point.

SEED KEEPER 1: "For me, seeds are my children. I protect them with the same love and the same care." (Navdanya, 2012)

SEED KEEPER 2: "Without seed there is no food. Without food there is no life. Saving seeds is saving life in all its diversity." (Navdanya, 2012).

SEED KEEPER 3: "The relationship between the farmer and seed is that between mother and son. These seeds that I am wearing[13] talk to me all the time. They say, 'Though I am small put me in the soil and I will do great work!'" (Navdanya 2012)

SEED KEEPER 4: "The seed is Brahma, the Creator. If it is living seed and fertile seed then we will have fertile life." (Navdanya, 2012)

SEED KEEPER 5: "We kept the traditional seed for so many generations; seeds from so many villages. Even those people who had little food, they never touched their seeds. Even if they starved they never ate their seeds. Seeds were kept separately." (Navdanya, 2012)

The ways in which seed keepers view the seed and care for the seed reflects the care they have for their own communities and nature. It is starkly different from the ways in which large corporations view seed as a commodity. Seed keepers are keepers of seeds, keepers of knowledge and keepers of communities. They are no longer simply farmers who are directed by large corporations and the market on what to grow, how to grow and for whom to grow. By taking are, they take control of their and their communities' regenerative capacity. They are motivated by the needs of their own community's production *and* reproduction. Using the term seed keepers helps us think of farmers as not just labor in the process of production, but as a source of reproduction through knowledge preservation.

Max Haiven's notion of "commons" helps us see the various campaigns by Navdanya around the maintenance of seed sovereignty and biodiversity as experiments in reimagining labor's relationship to nature and its position in the productive collective. Haiven (2014) describes commons as "existing alternative anti-capitalist institutions that make life worth living" (p. 23). He further explains, "in this sense, we must think of 'common' as a verb and not a noun" (p. 23). In doing so, he says, "we 'common' all the time, whenever we build the structures of relationality, solidarity, and cooperation that make up our lives, from our most private moments to our most public engagements" (Haiven, 2014). Haiven then goes on to explain that "memory is a commons that we draw on and contribute to in the present, and that 'history' (as a set of 'finished', authoritative stories) is a form of enclosure." By reimagining commons, Haiven asserts that memory can be an important countervailing force to capital's constant attempt to influence how we common.

Farmers not only create a common space for the physical preservation of seeds but also an intangible network in the local community that preserves traditional knowledge. This preservation of traditional knowledge takes the form of collective memory and resists capitalism's attempts to discard and destroy traditional forms of farming. Caffentzis and Federici (2014) further contend: commoning initiatives are more than dikes against the neo-liberal assault on our livelihood. They are the seeds, the embryonic form of an alternative mode of production in the make" (p. i95). What better place to start commoning than the seed-the starting point of life.

The focus on seed preservation for local communities also shifts the focus from just production (measured in yield per acre) of food to conversations around nutrition and resilience. Most seed keepers are interested in preserving the seeds that provide nourishment to their communities. They are interested in seeds that are resilient to the local climate so that the process of reproduction is not hindered due to natural calamities. This is particularly relevant in the face of India's nutrition crisis. According to the Global Hunger Index (2021), India ranks 101 out of 121 countries in world hunger with 15.3% of the population undernourished.[14] Undernourishment is generational, according to Jean Drèze (2019), who in an interview with Rathmore for the *Diplomatist* said, "Eradicating malnutrition is bound to take time, because it has an intergenerational aspect. When children are underweight at birth because their mothers are undernourished, it is difficult to make up for that deficit later on" (n.p.).

As such having access to nutritious food is very important for the reproduction of communities. In this vein, Navdanya has questioned the use of terms like "yield per acre" in traditional economics arguing that what is important it to measure "nutrition per acre." According to Navdanya, monoculture and excessive use of fertilizers and pesticides has led to a crisis of nutrition. During our visit to the biodiversity farm, the resident soil scientist explained to us how the excessive use of pesticides had led to a drastic decrease in the number of earthworms and other such living organisms that contribute to the nitrogen production in soil. The move to the use of seeds that do not require excessive chemicals, is therefore not only necessary for nutritious food but also for the preservation of soil for future generations.

As mentioned earlier, the seed keepers at Navdanya are mostly women. According to Caffentzis and Federici (2014), women's precarious relationship with wage employment has always made them more in tune with defending nature's common resources. Shiva (2014), pointing out to the role of women in the ecology movement, further explains that

> By starting a partnership with nature in the politics of regeneration, women are simultaneously reclaiming their own and nature's activity and creativity. There is nothing essentialist about this politics because it is, in fact, based on denying the patriarchal definition of passivity as the essence of women and nature. There is nothing absolutist about it because the natural is constructed through diverse relationships in diverse settings. Natural agriculture and natural childbirth involve human creativity and sensitivity of the highest order, a creativity and knowledge emerging from partnership and participation, not separation. The politics of partnership with nature, as it is being shaped in the everyday lives of women and communities, is a politics of rebuilding connections, and of regeneration through dynamism and diversity.
>
> (n.p.)

The recentering of women in food production at Navdanya recenters the act of reproduction in a powerful way.

Navdanya's efforts to save and protect seeds is often referred to as "seed satyagraha." Satyagraha, which translates into "force of the truth", is a term that was used by Gandhi during India's independence struggle, when he refused to follow unjust laws. Navdanya's use of this term is deliberate as it reminds us that the struggle against colonization and imperialism continues. It is also a non-violent approach to resistance thereby adhering to the successful tactic adopted by Gandhi for India's independence.

As mentioned earlier, the seed banks managed by the seed keepers "loan" seeds to local farmers, who are expected to return the seeds with interest. This use of the word's "banks" and "interest", we believe is an intentional and savvy use of language. The bank in our mainstream understanding refers to highly financialized institutions that often engage in predatory loans. The 2007–2008 crisis is one example of how the recklessness of banks led to an economic collapse. These banks then had to be bailed out by government because they were "too big to fail."

Navdanya's seed banks on the other hand are providing sources of reproduction to impoverished communities. They are not predatory but nurturing.

During our own visit to the biodiversity farm, while showing us the seed bank, our guide referred to it as a common temple for all that is more important than any religious place of worship.

The banks' physical manifestations also make them a common resource. For example, a farmer associated with Navdanya says, "We keep our own seeds and store wherever there is a space, under the cot, in the window on the roof" (Navdanya, 2012). In a village in Tehri Garhwal, seed keepers use their own house as seed banks. They share these seeds not only with their fellow villagers but also with farmers from other villages. A traditional bank separates us from our source of reproduction. In other words, it's a separate geographical (now more of a separate digital) space where our savings, our futures are held hostage to the economic uncertainty of the capitalist system that we live in.[15] Navdanya's seed banks on the other hand are embedded in people's homes, their spaces and their lives. They have complete access and control over their source of reproduction.

The decentralized and common location of the seed banks ensures that a large number of communities have access to seeds. The seed banks store seeds that are local to the agro-climactic conditions of that region. It therefore allows farmers to grow crops that tend to be more resilient. Contrast this to the universalizing (dystopian) approach to seed preservation that we see in Svalbard called "The Vault," which is located in the far end of the world protecting seeds for the future in the case of an environmental collapse. "The Vault" represents a stock waiting to be utilized when a disaster strikes, and Navdanya's community seed banks on the other hand represent a continuous regenerative flow that is working to stop such a disaster.

The word "interest" is also Deliberate: Interest in terms of seeds signifies life, it signifies the preservation of future. This is in complete contrast to the word's use in mainstream banking, where interest is an extractive form of economic exchange. This notion of seeds as interest creates a different relationship between "lender" and "borrower." This relationship is built around generosity and contingency. The repayment of the loan and interest is not controlled by market forces and economic reality but is rather based on the reality of their own lives. In case there is a crop failure, farmers are not expected to return their loans and interests but are provided with more seeds. This relationship also frees farmers from the debt relationship that they have with their money lenders.

As we hope this chapter has made clear, Navdanya helps us envision spaces that are not dominated by capitalist relationships. It provides us an example of what Gibson-Graham calls, "the creative forces and subjects of economic experimentation" in order to create the radical grounds for a "lived project of socialist construction" (2006, p. 251). As Gibson-Graham, Cameron and Healey (2013) assert, the act of reframing plays a crucial role in the transition to a sustainable and just society. In addition to providing an alternate language to resist and reject capitalism, Navdanya reimagines post capitalist subjectivities in the form of the seed keepers through a language of care and nuturing.

As with each of our sites, Navdanya perfectly illustrates Susan George's (2004) assertion that:

> If neither mass personal transformation nor one-off revolutionary change can be counted on to create another world, perhaps we need to be a bit more modest. If capitalism does one day suffer defeat, I believe it will be as the cumulative result of hundreds of struggles, not of some great global apocalypse. So I humbly suggest that we just have to keep on working to change the balance of forces in the most imaginative and unviolent ways we can find, while keeping an eye out for what history has on offer.
>
> (p. 96)

Notes

1 The popular term "tree hugger" emerged from the Chipko movement.
2 See Nelson, Ravichandran and Antony (2019) for further discussion on the loss of indigenous varieties of crops due to the Green Revolution.
3 See report by Center for food safety (2015) for a description and discussion on the growing influence of multinational corporations like Monsanto in the seed industry.
4 See Center for Human Rights and Global Justice (2011) for further description and discussion of the farmer's suicide crisis in the cotton growing regions in India. https://chrgj.org/wp-content/uploads/2016/09/Farmer-Suicides.pdf
5 Information recovered from visit and conversation with staff at Navdanya Biodiversity farm.
6 Based on conversation with the seed keeper expert during our visit to Navdanya Biodiversity farm.
7 Zamindars were the landlord class in the colonial period in India.
8 After more than 40 years of planned economy, India opened up its economy to the world markets in 1991. Liberalization in India was a response to the payment crisis, which was faced by the central government in 1991. The main reason for the payment crisis was the sharp increase in oil prices, which led to a trade deficit.
9 The bills were rushed through the parliament without much deliberation.
10 The repeal of these laws by the government was mainly due to the forthcoming state elections in Punjab that has a large farming community and therefore is an influential voting bloc.
11 The ADC was founded by J.D. Rockefeller, III in 1953 to train agricultural economists who would then go back to their countries to formulate agricultural policies
12 There are a number of organizations that are committed to saving seeds like Basudha in the state of Orissa, Sahaja Samrudha, Save Our Rice Campaign, Kheti Vriasat, Sampark etc.

13 He says wearing seeds because he carries different varieties of seeds on his hat and takes them to different local fairs and exhibitions.
14 Undernutrition refers to deficiencies in energy, protein, essential vitamins and minerals (Global Hunger Index).
15 The negative interest rate regime in Switzerland led to increased fees for bank account users. In other words, customers were required to pay a fee to keep their money in banks.

References

Ajl, M. and Sharma, D. (2021) The Green Revolution and transversal countermovements: Recovering alternative agronomic imaginaries in Tunisia and India. *Canadian Journal of Development Studies.* doi: 10.1080/02255189.2022.2052028

Alavi, H. (1975) India and the colonial mode of production. *Economic and Political Weekly*, 10(33–35), pp. 1235–1262.

Anderson, R. and Morrison, B. (1982) Introduction in science, politics and the agricultural revolution in Asia. In Anderson, R., Brass, P., Levy, E. and Morrison, B. (eds.). *Science, politics and the agricultural revolution in Asia.* New York: Routledge, pp. 1–12.

Bagchi, A. (1982) *The political economy of underdevelopment*. Cambridge: Cambridge University Press.

Bhaduri, A. (1983) *The economic structure of backward agriculture.* London: Academic Press.

Bhalla, S. (1989) Employment in Indian agriculture: Retrospect and prospect. *Social Scientist*, 17(192–93), pp. 3–21.

Caffentzis, G and Federici, S. (2014) Commons against and beyond capitalism. *Community Development Journal*, 49(51), pp. 92–105.

Center for Human Rights and Global Justice (2011) Every thirty minutes: Farmers suicides, human rights, and the agrarian crisis in India. New York University School of Law. Available at: https://chrgj.org/wp-content/uploads/2016/09/Farmer-Suicides.pdf

Center for Food Safety and Save Our Seeds (2013) "Seed Giants vs. U.S. farmers." Retrieved from www.centerforfoodsafety.org/files/seed-giants_final_04424.pdf

Chakravarty, S (1987) *Development planning: The Indian experience*. Oxford: Clarendon Press.

Cleaver, H.M. (1972) The contradictions of the Green Revolution. *The American Economic Review*, 62(1/2), pp. 177–186.

Cleaver, H.M. (1982) Technology as political weaponry. In Anderson, R., Brass, P., Levy, E. and Morrison, B. (eds.). *Science, politics and the agricultural revolution in Asia.* New York: Routledge, pp. 261–275.

Drèze, J. (2019) Interview with Rathore, H. Jean Drèze on malnutrition and child mortality in India. Retrieved from https://thediplomat.com/2019/12/jean-dreze-on-malnutrition-and-child-mortality-in-india/

George, S. (2004) Another world is possible if… London: Verso.

Gibson-Graham, J.K. (2006) *The end of capitalism (as we knew it): A feminist critique of political economy*. Minneapolis: University of Minnesota Press.

Gibson-Graham, J.K., Cameron, J. and Healy, S. (2013) Take back the economy: An ethical guide for transforming our communities. Minneapolis: University of Minnesota Press

Global Hunger Index (2021) Global Hunger Index-Hunger and food systems in conflict settings. Available at:www.globalhungerindex.org/pdf/en/2021.pdf

Haiven, M. (2014) Crises of Imagination, Crises of Power: Capitalism, creativity and the commons. London: Zed Books.

Harvey, D. (2005) A brief history of neoliberalism, Oxford University Press.

Nadkarni, M. V. (1996) Forests, people and economics. Indian Journal of Agricultural Economics, 51(1&2), pp. 1–24.

Navdanya (n.d.) *Conserving diversity and reclaiming commons.* Available at: www.navdanya.org/site/latest-news-at-navdanya/705-gardens-of-hope

Navdanya (2012) *Seed freedom: A global citizens' report. RS (Coordinators).* India: Navdanya, October.

Navdanya (2016) *Navdanya partners with VNSSM and CICR to make Maharashtra suicide, poison and debt free.* Available at: https://seedfreedom.info/navdanyas-partners-with-vnssm-and-cicr-to-make-maharashtra-suicide-poison-and-debt-free/

Navdanya (2020) *Navdanya.* [online video] Available at: www.youtube.com/watch?v=YPaZtaCN7_k

Nelson, E., Ravichandran, K. and Antony, U. (2019) The impact of the Green Revolution on indigenous crops of India. *Journal of Ethnic Food*, 6(8). https://doi.org/10.1186/s42779-019-0011-9

Osmani, S.R. (1993) Growth and Entitlements The Analytics of the Green Revolution. World Institute for Development Economic Research (UNU-WIDER), WIDER Working Papers.

Patnaik, P. (2020) Agriculture bills and food security. Available at: www.networkideas.org/news-analysis/2020/10/agriculture-bills-food-security/

Patnaik, U. (2004) It is a crisis rooted in economics reforms: Interview with Professor Utsa Patnaik. *Frontline*, 21(13). https://frontline.thehindu.com/cover-story/article30223320.ece

Seed Freedom (n.d.) *Bija Devi—Navdanya's seed keeper of 16 years at the biodiversity conservation farm.* Available at: https://seedfreedom.info/staff/bija-devi-navdanyas-seed-keeper-of-16-years-at-the-biodiversity-conservation-farm/

Seed Freedom (2020) *Navdanya Seeds—Naha Raj Singh.* [online video] Available at: www.youtube.com/watch?v=7rFndP-HOy0

Sen, A.K. (1959) The choice of agricultural techniques in underdeveloped countries. *Economic Development and Cultural Change*, 7(3, Part 1), pp. 279–285.

Sen, A. and Ghosh, J. (2017) Indian agriculture after liberalisation. *The Bangladesh Studies*, Special Issue in honour of Manabub Hossain Agricultural Tranformation, Structural Change and Policy reforms, 40(1 & 2), pp. 53–71.

Shiva, V. and Bandyopadhyay, J. (1986) The evolution, structure, and impact of the Chipko movement. *Mountain Research and Development*, 6(2), pp. 133–142.

Shiva, V. (1991) *The violence of the Green Revolution Third World agriculture, ecology and politics.* London and New Jersey: Zed Books.

Shiva, V. (2014). *The Vandana Shiva reader.* Lexington: University Press of Kentucky.

Shiva, V. (2016) *Soil, not oil: Climate change, peak oil and food insecurity.* London: Bloomsbury Publishing.

The Tricontinental (2019) The neoliberal attack on rural India: Two reports by P. Sainath. *Tricontinental: Institute for Social Research*, 21, October edition.

Walker, L.K. (2008) Neoliberalism on the ground in rural India: Predatory growth, agrarian crisis, internal colonization, and the intensification of class struggle. *The Journal of Peasant Studies*, 35(4), pp. 557–620.

WTO (n.d.) Uruguay round agreements: TRIPS: Part II—Standards concerning the availability, scope and use of intellectual property rights, Section 5 and 6. Accessible at: www.wto.org/english/docs_e/legal_e/27-trips_04c_e.htm

6 "No justice, no java"[1]

Re-commoning the continent at Tonatierra

Tonatierra is an Indigenous owned and operated non-profit (501c3) organization that is guided by a "calpolli"[2] ("Nahuacalco") comprising 20 families in Phoenix, Arizona, in the southwestern region of the United States. "Calpolli" is the Nahuatl[3] ("large house") term to describe the main social, spatial and economic unit of pre-Colombian Aztec societies and which formed the "basic unit of community identity in the Mexica culture" (Tonatierra, 2022a). The founding families of Tonatierra revived this ancestral form of self-governance a generation ago in order to guide their work and reactivate "the path of tradition and liberation, resistance and regeneration" at the level of the *community* and the *everyday* (Tonatierra, 2022a, italics added). The name "Tonatierra" is a composite of the Nahuatl word "tona" meaning "sun" and Spanish word "tierra" meaning "earth" and is meant to highlight the interwoven histories and cultures of the traditional peoples of the Americas. According to the community webpage, the mission of Tonatierra is:

> To create and sustain a Cultural Embassy of the Indigenous Peoples that will support local-global and holistic indigenous community development initiatives in education, culture, and economic development in accord with the principles of Community Ecology and Self Determination.
>
> (Tonatierra, 2022b)

Tonatierra comprises two separate but related parts: the non-profit (501c3) and the community organization, both of which are organized from the "Nahuacalli" ("the house of the four directions"), a building and site owned by the organization in downtown Phoenix. Community member and "Calpixqui" ("housekeeper" or "gatekeeper"), Eve Reyes-Aguirre describes Tonatierra as a "vehicle" to do the traditional work of the community and the Nahuacalli as a "home" or "embassy" where members of the Indigenous community can come together in their common struggles against the forces of extractivism and in the pursuit of self-determination and cultural regeneration.

Tonatierra was born in the ferment of the 1990s—a period characterized by the rekindling and reconnection of Indigenous cultures and communities and the simultaneous (and reactionary) intensification of neoliberal,

DOI: 10.4324/9781003138617-6

Figure 6.1 The Nahuacalli Cultural Embassy.

neocolonial programs of enclosure and exploitation.[4] These contending forces were perhaps nowhere more volatile than in the Americas.[5] Reyes-Aguirre describes this moment, and the wider Indigenous movement, as a rebirth of "who we are as a people and how we lived culturally and economically" (Reyes-Aguirre, June 7, 2022). In the Americas, the 1990s saw the birth of the Continental Indigenous Movement bringing together Indigenous communities in "Encounters" and "Summits" around the shared experiences of how colonialism affects Indigenous communities and shared visions of self-determination. Founders, and present-day elders, of what would become Tonatierra partook in these continental gatherings, sowing the seeds for its formation in response to the 1994 signing of NAFTA and consequent Zapatista Uprising in Chiapas Mexico. Community members "returned to Chiapas as Indigenous Human Rights observers with the intent of opening venues for non-violent resolution of the issues in conflict and in solidarity as Indigenous Peoples" and came home to Phoenix with a mission to sell local farmers' coffee: the "Quetzal Co-op" and the "Nahuacalli" cultural embassy answered this call. Since its inception, Tonatierra has been oriented to "fulfilling their obligation to the movement" and in this vein adapts its programs, projects and mission to meet the needs of the local, regional, global community as they arise (Reyes-Aguirre, 2021). The calpolli as an organizational unit is central to their ability to do this work. In addition to their involvement in local and regional advocacy and social justice work,[6] and in the spirit of revival and regeneration, the community is committed to repairing the psychological and cultural damage done by colonialism. As the website states:

> for over 522 years, the Original Nations of the Great Turtle Island Abya Yala have endured one of the most extreme forms of colonization and genocide in all of human history. The Program of genocide has extended into the very fabric of our conceptions, intellectual processes, historical self-awareness, and psychological health.
>
> (Huehuecoyoytl, 2022)

Cultural repair and regeneration is especially oriented toward the community's children.

Figure 6.2 Children of Tonatierra on a mural.

The community meets every 20 days to discuss the various educational, cultural and advocacy programs and projects. Decisions are made by consensus: in the case of disagreement, it is the responsibility of the dissenter to convince the group of their differing position; if no agreement is arrived at, the group defers to community elders for a decision. The community is guided informally by elders who, in addition to assisting in matters of disagreement, advise on spiritual, cultural and historical matters. Women play as important a role as men in leadership and decision making. The day-to-day decision making on the matters involving the non-profit are made by the "Tenamaztle" (translated as "three stones of the hearth" and loosely meaning "foundation").

This chapter explores Tonatierra as a culturally specific, place-based approach to the larger project of reimagining economic relations and subjectivities representative of our contemporary moment. The next section situates Tonatierra in the broader colonial and neoliberal history of North America and its representation in the struggle over and around the US–Mexico border.

"The border crossed us"

The longtime battle cry of the Mexican/Chicanx immigrant rights movement, "we did not cross the border, the border crossed us," seems an appropriate frame for approaching Tonatierra—as a community, as a site and as a struggle. Tonatierra was born of a desire to repair and reconnect the divisions and disruptions caused by 500 plus years of colonial exploitation, extraction, extermination and displacement. Arizona is central in this history. While it is outside the bounds of this essay (as well as our professional expertise and personal experience) to detail the colonial history and present of Arizona, we want to acknowledge the region's significance as a literal and symbolic "borderland"[7] and site of struggle. The "borders" that constitute present-day Arizona have been drawn and redrawn over the past 500 years, and the state's population, culture and conflicts are representative of these changes: From the borderless ancestral *lands*[8] of pre-colonial peoples (Ancestral Puebloans, Hohokam, Mogollon, Sinagua, Salado, Cohonina and Patayan as well as the nomadic Apache and Navajo); to a site of Spanish *settlement* in the sixteenth and seventeenth centuries; to a *region in the state* of Sonora, Mexico from 1822–1848; to its partial possession in 1850,[9] complete seizure ("purchase") in 1853,[10] and eventual designation as a US *colonial territory* in 1863 with the gradual addition of Native American *reservations* over the next century;[11] to its absorption as a US *state* in 1921; to its contemporary articulation as a *militarized fortification* (Sheridan, 2012; Sheridan and McGuire, 2019). Far from being mere lines on paper, as Arizona poet laureate Alberto Ríos puts it, "those lines are our lives" (2007): They are drawn through and around and against individuals, families, communities, habitats and histories. His 2007 poem, *Líneas fronterizas/Border lines*,[12] challenges us to shift the "way we look" at those lines:

A weight carried by two
Weighs only half as much.

The world on a map looks like the drawing of a cow
In a butcher's shop, all those lines showing
Where to cut.

That drawing of the cow is also a jigsaw puzzle,
Showing just as much how very well
All the strange parts fit together.

Which way we look at the drawing
Makes all the difference.
We seem to live in a world of maps:

But in truth we live in a world made
Not of paper and ink but of people.
Those lines are our lives. Together,

Let us turn the map until we see clearly:
The border is what joins us,
Not what separates us.

In addition to drawing attention to the casual violence of maps as they naturalize partition, Ríos's poem also reminds us that the violence they intimate is a result of perspective. Viewed a different way, borders are seams not tears.

Ríos's perspective is informed by his own upbringing in the "borderlands" of the Mexico–US border in Nogales, Arizona, just south of Phoenix. The twin towns of Nogales (Arizona and Sonora) came into existence when the first railroad lines connected the US and Mexico in 1883. For 46 years, the two towns functioned as one until it was "cleaved by a continuous barrier in the form of a two-meter-high chain-link fence erected in 1929" (McGuire and Van Dyke, 2019, p. 45). While there were minor alterations and additions to the fence in the years between 1929 and 1994, the "relatively permeable, neighborly chain-link barrier defined the border and symbolized Ambos Nogales's neighborly relationship" (ibid., p. 45). McGuire and Van Dyke (2019) describe the pre-NAFTA Nogales as a space of relative cultural hybridity and social and economic interdependence: Celebrations transcended the border with parades of bands and floats passing through the main gate; the economies of both cities depended on the border; and civic leaders on both sides had binational and bicultural knowledge, orientations, and social networks.

This articulation of the border changed abruptly when, "concurrent with the implementation of NAFTA in 1994, the United States adopted a 'policy of deterrence' to stop unruly bodies (undocumented migrants and drug smugglers) from crossing the border" (2019, p. 46). In order to force unwanted migrants to cross in the desert, the US government attempted to make crossing at the border city difficult by replacing the permeable chain-link fence with a taller "wall constructed of military surplus landing mats topped with an angled steel anti-climb guard" (ibid., p. 46). Following the September 11, 2001 terrorist attacks in New York and Washington DC, the US government increased border security by "adding multiple human, animal, and technological layers" to the walls (ibid., p. 46). The militarization of the US–Mexico border escalated with the formation of ICE (Immigration and Customs Enforcement), a US federal law enforcement agency in 2002 and the erection of newer, higher walls in border towns in the US southwest. While these attempts to slow migration have failed, they have, on the other hand, seen the incredible rise of violent confrontations at the border and tragic deaths in the desert (Haddal, 2010). These failures point to the paradox of the neoliberal era as one of increased cross-border flows (of goods, capital, culture, ideas and some people) and the simultaneous implementation of new forms of exclusion, privilege and marginalization that disproportionately affect the poor and powerless.

Free trade and captive borders

> The border used to be an actual place, but now, it is the act of a thousand imaginations.
>
> (Alberto Ríos, The border: A double sonnet)

For much of contemporary history, migration across the US–Mexico border has ebbed and flowed; its enforcement tied to the demands of economic and political elites (mostly in the United States). The militarized fortification that now divides the US–Mexico border is evocative of the contemporary "worldwide wall-building movement" that has seen the proliferation and weaponization of borders (McAtackney and McGuire, 2020). This new "border regime" is the complex result of the "neoliberal counterrevolution," which has seen the development of "two nested spheres—the planetary networks of social production and reproduction, and the constitution of global governance—that are increasingly out of sync" (Hardt and Negri, 2019, p. 68). Borders take on a special significance in this system as sites of symbolic struggle and "imagination" as Ríos suggests. These "border regimes" have their prehistory in the postwar period: a moment of radical economic reform and equally radical social activism in a rapidly (and unevenly) globalizing world. In the context of Mexico and the US, this period saw the rapid transformation of traditional economic relations that favored US corporations and agricultural producers and disrupted labor patterns on both sides of the border.

According to McCaughgan (2020), in this period Mexicans and Mexican Americans

> experienced the inhumane vicissitudes of US immigration policies aimed at alternately shrinking and enlarging the labor pool to meet the needs of

> an economy that contracted dramatically during the Great Depression and then accelerated during and following World War II.
>
> (p. 10)

The infamous "Braceros Program"—a series of diplomatic accords between Mexico and the United States from 1942–1964 addressing agricultural labor shortages during and following WWII—temporarily "imported" more than four million "Braceros" ("strong-armed ones") across the border to work in agriculture. The program's curious format as a "one-nation (Mexico), one-sex (male), one sector (agriculture), and one season at a time" immigration program hints (more like shouts) at its view of the laboring man as a disembodied, disembedded "arm" that could be rented and returned at the whims of the needs of a capital class (North, 2013, n.p.) Because "Braceros" became the choice laborers of growers, wages stagnated for domestic farm workers (for why hire entire bodies with rights when you could hire arms with none …?). What is more, "once the growers became used to these conditions they, of course, regarded the cheap labor situation as their God-given right, and used their political prowess to preserve these privileges" (North, 2013, n.p.). The financial exploitation and physical abuse of the Braceros (built into the program because visa were tied to employers)[13] saw the development of the farm workers movement (and the eventual creation of the United Farm Workers union under the leadership of César Chávez and Dolores Huerta). While the movement had many important consequences for labor, it also saw the retaliatory and violent mass deportation of millions of Mexicans and Mexican Americans (McCaughan, 2020).

Though the agreement formally ended in 1964 (in the context of growing civil unrest and union pressure), the 20-plus year migration of farm workers to the north had enduring impacts on both sides of the border. The gaps in food production left by Mexican farm laborers were increasingly filled by US American agricultural imports (sedimented with the 1994 implementation of NAFTA). What is more, the end of Braceros program did not end the uses or abuses of migrant labor for, as North (2013) attests, many growers simply switched to "illegal" migrant workers as there was not enough interior immigration law enforcement to "get many growers to adopt modern labor-management practices. Western growers, particularly, resisted the mild regulations that went with (and go with) the H-2A [temporary agricultural] program" (n.p.). There is also no denying the relationship between the farm labor movement and the Chicano movement that began in the same period as a "constellation of civil rights struggles by Mexican American communities on many different fronts, including educational institutions, agricultural fields, and urban neighborhoods" (MacCaughan, 2020, p. 17). In addition to addressing the borderland issues of racism, cultural assimilation and expression (e.g. language, art), police brutality, immigration, and labor issues, the movement also expressed solidarity with decolonial struggles (MacCaughan, 2020).[14]

The Bracero program was immediately followed by the Mexican Border Industrialization program (BIP) in 1965. While Mexico, like many other developing nations in that period, was committed to the path of Import Substitution Industrialization in order to lessen its reliance on foreign imports, the BIP allowed US companies to set up factories in a 20 km strip on the US–Mexico border (Buchanan, 1995). The program was pitched as a solution to the "excess labor" created by the ending of the "Braceros" program; in reality, however, the factories employed mostly unmarried, young women whose jobs were highly segmented, low paid and low skilled (Buchanan, 1995). They were also largely non-unionized resulting in substandard working and living conditions. Buchanan referred to the US–Mexico borderland of this period as a "region of 'hope and heartbreak' and 'boom and despair'" (p. 383).

The Mexican economy was heavily impacted by the global economic crisis of 1970 with the 1980s often referred to as the "Lost Decade." In response to the impacts of the crisis in Mexico, and in order to stabilize employment, the government, under the Partido Revolucionario Institutucional (PRI), nationalized several failing private companies with loans taken from banks in the United States. The 1979 "Volker shocks" increasing interest rates in the US in an attempt to combat inflation there, caused Mexico to default on their loans and led to the Mexican debt crisis of 1982. The World Bank, the IMF and the US Treasury stepped in at this time offering conditional loans requiring a series of neoliberal structural adjustment programs including the privatization of the state held companies (that had been nationalized using loans from the US ...), the minimization of the power of labor unions in contract negotiations, and the further opening of its economy to international trade[15] (Harvey, 2005; Flores-Macías and Sánches-Talanquer, 2019). Negotiations for a regional trade agreement began shortly thereafter in 1990 between Mexican President Carlos Salinas and US President George H.W. Bush with Canada joining the negotiations in 1991. The result, the North American Free Trade Agreement (NAFTA), was implemented on January 1, 1994. The agreement gradually eliminated tariffs and non-tariff barriers (e.g. quotas and export subsidies). A few months earlier in September 1993, and in the context of concerns by the US and Canada that Mexican labor and environmental standards would not be on par with those required by Canadian and US law, the three countries signed the North American Agreement on Labor Cooperation (NAALC) and the North American Agreement on Environmental Cooperation (NAAEC).

The most serious impacts of NAFTA on Mexican people were in the agricultural sector: Given the economies of scale of American agri-businesses and their continued subsidization (mainly corn and wheat) following NAFTA, US agri-businesses could sell their products at a much lower price than local producers in Mexico. Many small-scale farmers (farming predominantly corn) as well as larger ranching and farming enterprises were unable to compete with the large firms from the US and had to give up farming and move to other regions for industrial and informal work (Grammont, Mackinlay and Stoller, 2009). Many of these jobs were precarious as they were connected to the emerging global supply chain and many were lost to China when it opened to international markets by joining the WTO in 2000. This period also saw the elimination of the *ejido* system (in Salinas's euphemistic "ejido reforms"), which had been established in the wake of the Mexican revolution (1910–1920), in order to protect the rights of Indigenous people to collectively hold and use their lands. According to Hasketh (2019), this system allowed for the establishment of communal forms of land tenure and an economy of subsistence farming; their eradication was devastating for Indigenous communities.

NAFTA also influenced south–north migration as US corporations built factories (mostly) in the northern industrial belts of Mexico; this resulted in the loss of US manufacturing jobs to Mexico and the migration of Mexican workers to these regions. According to Buchanan (1995, p. 387), this inevitably prompted cross border migration for, as she points out, "migration to the border and migration across it are closely linked phenomena". While Mexico attempted to bring border migration into the agreement, these requests were ignored and, to the contrary, the US began militarizing its borders immediately following NAFTA's signing. Inspite of the border's reinforcement, the number of undocumented people in the US has tripled since NAFTA was put into effect (Verea, 2014).

"You build a wall, we build a bridge"[16]

> We have come to realize that we are not alone in our struggles nor separate nor autonomous but that we [...] are connected and interdependent. We are each accountable for what is happening down the street, south of the border or across the sea.
>
> (Anzáldua, This Bridge Called My Back, 1981)

While there were many localized attempts[17] by rural and Indigenous groups to address the manifold issues brought on by Mexico's radical economic changes in the 1990s (especially as they affected rural and Indigenous peoples), they had limited successes until the Ejército Zapatista de Liberación (EZLN) put neoliberalism (and its discontents and benefactors) into the international spotlight with their seizure of six towns in the Chiapas on January 1, 1994. The Zapatista Uprising or Insurgency was a carefully coordinated occupation-turned-media-spectacle that began on the day that the North American Free Trade Agreement (NAFTA) took effect. The Zapatista's declaration of war was the product of years of (suppressed) efforts to politically and socially mobilize. The Ejército Zapatista de Liberación (EZLN) began as one of the many small, armed militant groups set up in response to the Mexican government's attempts to suppress Indigenous mobilization. In 1993, the EZLN designated Subcomandante Marcos their ideological leader.

This was unusual because Marcos is "mestizo." According to Harvey (1995), the Zapatisa rebellion was

> unlike previous guerrilla struggles in Latin America in that it [did] not aspire by itself to seize state power and lead the masses in social revolution. In its declaration of war on the federal army and government, the Zapatistas called on all Mexicans to participate in whatever way they can—not necessarily with arms—in a broad movement for "jobs, land, housing, food, health, education, independence, freedom, democracy, justice and peace."
>
> (p. 39)

The main demands of EZLN were the repeal of Salinas's "*ejido* reforms" of 1992 and the redistribution of the latifundios (Harvey, 1995).

According to Goldelmenn (2014), the Zapatista movement was instrumental in bringing the plight of the Indigenous peoples of southern Mexico to national and international attention. While a ceasefire was called by the Mexican government on January 12, a peace agreement was not agreed upon until 1996, with the signing of the San Andres Accords (Solano, 2007) which gave some autonomy to Indigenous peoples in the Chiapas. In 2001, the Mexican Congress also passed a law recognizing the multi-cultural composition of Mexican society and gave Indigenous peoples the right to determine their political status and to pursue their own cultural, economic and social development (Goldelmenn, 2014). According to Bréton et al. (2022), the Zapatista movement has managed to maintain some control over the region of Chiapas as they have "pursued their project of autonomy based on reinforcing subsistence production and indigenous self-government" (p. 568).

The rise of the EZLN cannot be seen in isolation of other forms of national and international resistance. Beginning in 1990, and in the spirit of building regional solidarity, Indigenous communities in the Americas began organizing a series of Continental Indigenous Encounters (1991 in Ecuador, 1993 in Mexico) and Continental Indigenous Summits (2000 in Mexico, 2004 in Ecuador, 2007 in Guatemala, 2009 in Peru, 2013 in Colombia). In addition to bringing diverse Indigenous communities together in the common experience of colonization and pursuit of self-determination, these encounters facilitated enduring dialogues and relations between Indigenous communities of the north and south. There is no denying the role of these summits (and the north–south relations they inspired) in the formation and global mediation of the Zapatista Uprising.

The Zapatista Uprising should also be read against the backdrop of the longer arc of the international Indigenous Rights Movement that came into full expression in the postwar period. According to Niezen (2003), four specific aspects of the postwar era contributed to a new climate in international politics that encouraged the promotion of Indigenous rights: First, the struggle against fascism contributed to a greater awareness among international bodies (mostly prominently the UN and also the ILO)[18] of the need for the protection of minorities against racism and discrimination; second, the dismantling of the European colonies raised awareness of the normalized forms of cultural suppression and oppression that had seemed a natural part of the "civilizing" process in earlier generations; third, assimilation policies that used formal education as a means of eliminating Indigenous culture and language and integrating Indigenous peoples into mainstream societies had failed in their goal of eliminating all vestiges of attachment to tradition, while "unintentionally contributing to intertribal identity, broader political unity, and the training of educated leaders" (2003, p. 8); and fourth, the proliferation of NGOs, which saw the constitution of an international civil society and provided the opportunity for Indigenous groups to organize and mobilize, even in nations hostile to their claims.

The 1970s–1980s saw the first inclusions of Indigenous issues in UN programs. The "Decade for Action to Combat Racism and Racial Discrimination" from 1973–1982 instigated the creation of various conferences, committees, working groups and delegations. The report by the Legal Commission summarizing the findings of a 1981 NGO conference at the United Nations in Geneva on the issue of Indigenous peoples and land is particularly salient:

> The dispossession of indigenous people from their lands and policies of forced assimilation have led to … untold social misery. Restoration of indigenous land base and agrarian reforms which would transfer the ownership of the land back into the hands of indigenous peoples without a requirement of either purchase or taxation are crucial.
>
> (United Nations 1981, p. 15)

While these plainly stated offenses and reparations have yet to be even remotely addressed anywhere in the world, international efforts culminated in the adoption of the United Nations Declaration on the Rights of Indigenous Peoples (UNDRIP) by the General Assembly on September 13, 2007, by a majority of 144 states in favor, 4 against (Australia, Canada, New Zealand and the United States) and 11 abstentions (Azerbaijan, Bangladesh, Bhutan, Burundi, Colombia, Georgia, Kenya, Nigeria, Russian Federation, Samoa and Ukraine). All four nations that voted against the declaration originated as settler colonies of the British Empire. Since 2007, each has endorsed the declaration in some informal way that would not be legally binding. Though unenforceable, the UNDRIP

> establishes a universal framework of minimum standards for the survival, dignity and well-being of the indigenous peoples of the world and it elaborates on existing human rights standards and fundamental freedoms as they apply to the specific situation of indigenous peoples.
>
> (UN, 2007)

While these various international movements have arguably had limited effects in changing the material conditions of Indigenous peoples, there is no underestimating the consequences they have had on building solidarity and community around Indigeneity and in creating languages and frameworks for thinking about restitution. In line with Rebecca Solnit's hopeful perspective on social movements and their long and often invisible arcs, the Zapatista Uprising was the product of countless, nameless individual and community efforts, which have in turn catalyzed countless, nameless more. Tonatierra is a part of this arc.

"The border is a line that birds cannot see"[19]: The Quetzal Co-op Project

The quetzal (Nahuatl for "standing tail feather") is a migratory bird species found in southern Mexico, Central and South America. It is the national bird (and currency) of Guatemala and is a sacred symbol of freedom in Mayan and Aztec mythology. According to the International Union for the Conservation of Nature, the "resplendent quetzal" found in the Chiapas and Guatemala has entered "Near Threatened" status (IUCN, 2021). The Quetzal Co-op is an initiative that emerged when members of what would become Tonatierra travelled to the Chiapas during the conflicts described above and offered their support. In response to their offers, local farmers in the Chiapas asked their allies in the north to buy their coffee; and so began the Quetzal Co-op (and many other similar partnerships). The Co-op started in 1996 as an initiative to support the "threatened status" of Indigenous coffee farmers in the Chiapas and has, in the years following, expanded to different regions in Mexico and Guatemala "buying green coffee beans from Indigenous coffee cooperatives above market value offering just trade prices and building economic and trade alliances from community to community, family to family" (Quetzal Co-op-Cafetzin, 2022). The coffee is roasted and packaged weekly (or as needed) in the Nahuacalli building by Tonatierra members. In addition to selling the coffee through their online marketplace, it is sold by several local vendors who are also committed to economic justice (see Quetzal Co-op, 2022). The proceeds from sales go back into the non-profit "to continue to build the cycle of support for the international economic networking of cooperatives of Indigenous peoples, north and south Abya Yala [Americas]" (Quetzal Co-op, 2022).

Like the quetzal bird and currency, the Quetzal Co-op is a migratory project built on the knowledge that the Americas were "a homeland in existence before modern borders turned Mexicans into immigrants and the continent's Indigenous communities into oppressed minorities in a colonized nation" and the concomitant objective to re-pair the economic relationships and practices severed by these processes (McCaughgan, 2020, p. 7). Far from viewing economic interactions as mere instrumental exchange (though they can be this too), the Quetzal Co-op is a clear example of an economic practice that is

Figure 6.3 Quetzal Co-op. Tombreck Farm Map (2022) Google, Geobasis-DE/BKG (© 2009).

"re-socializing the economy" by replacing the "asocial economic atomism" at the core of market-based exchanges with practices that recognize "economic co-implication, interdependency, and social connection" (Gibson-Graham, 2006, p. 83). While Quetzal Co-op partakes in fair trade networks and programs (as a means of connecting with growers and ensuring fair pricing), the project cannot be understood from the traditional lens of fair trade. As Naylor's (2018) reframing of fair trade from a "community economies" approach reveals, fair trade networks also facilitate other exchanges including those between roasters and growers. While the interdependencies between roasters and growers "are still mired in unbalanced privilege," (ibid., p. 1041), Naylor contends that the roaster–grower interdependencies are ones that "potentially go beyond the sale of a commodity, the creation of surplus value and the connecting of producer and consumer" (ibid., p. 1027) and, perhaps more significantly, are more likely to endure "if fair trade certification disappeared tomorrow" (ibid., p. 1041). This is due to the fact that the exchange is not mediated through a neutral, disembedded and disembodied market (like the producer–consumer exchange) but rather (and deliberately) through embodied negotiations that are meant to exchange coffee not as an end but as a part of a larger economic and political project of self-determination for Indigenous communities throughout the Americas. As stated on the Tonatierra webpage, Quetzal is:

> the articulation of a long range comprehensive economic development plan that originates from the Indigenous Peoples themselves to conduct COMMUNITY COMMERCE within an ethic of justice and mutual decolonization as we engage in the markets of goods and services from the local to regional, continental to global spheres of transaction.
>
> (Quetzal-Co-op, 2022)

Though the diverse participants have different needs, privileges and abilities in these arrangements, the economic exchange (on the part of Tonatierra) is based in the recognition that all struggles for Indigenous self-determination, north and south, are interconnected, as they state: "We work within a dynamic of establishing not just an economic relationship based on the exchange of coffee, but in realization of continental cultural solidarity as Indigenous Peoples" (Quetzal-Co-op, 2022). This does not stop at the grower–roaster exchange. As mentioned above, in addition to being sold online, Quetzal is sold by various community-oriented and ethical vendors who represent another node in the growing network of solidary. Tonatierra also sells Quetzal coffee in the various traveling "Tiankizcos" (Nahuatl for "marketplace") where they sell other justice-oriented goods.[20] They have also recently partnered with Café Sin Fronteras (another organization, in California, selling coffee from growers in Chiapas) in their "tip the farmer" program (see Quetzal-Co-op, 2022).

As with their other programs and projects, the Quetzal Co-op project is symbolic of the figure after which it is named: adaptive and responsive, migrating to different regions, communities and marketplaces as needed.

"Contemporary expressions of the ancestral way of life"[21]: The calpolli as commons

As described in the introduction to this chapter, Tonatierra is organized in the Mexica tradition of the "calpolli." Community "calipixqui" ("housekeeper" or "gatekeeper"), Eve Reyes-Aguirre, describes the Tonatierra "calpolli" as a form of community based in "kinship" as opposed to "membership" and constituted through traditional ceremonies and practices and relations of mutual accountability. Tonatierra is one of several resurgent calpolli in North America; Reyes-Aguirre cited other similar calpolli in Los Angeles, California, Tucson, Arizona, El Paso, Texas, and New Mexico, with whom Tonatierra has forged bonds of kinship "through travel, relations, and acquaintanceship" (Reyes-Aguirre, 2022). The community hopes to continue to build regional connections with these and other emergent calpolli.

The elders of Tonatierra revived this traditional form of self-governance in solidarity with local Indigenous farmers in Chiapas who rose up against the Mexican government in 1994 in response to the ongoing and radical changes to the economy symbolized and sedimented with the passage of NAFTA. A central target of the Zapatista's[22] critique was the privatization (enclosure) of the communal lands secured for Indigenous communities following the Mexican Revolution. In response to the reversal of the laws protecting these lands, the Zapatistas brought back the "calpolli" both as an organizational unit but also, and equally importantly, as a prefigurative embodiment of the world, not as it was, but as it should and could be (and indeed *had* been). In its most traditional sense the calpolli was a pre-colonial agrarian organizational structure that was connected to a piece of cultivated land. In Mexico, the calpolli would have been centered around the production of corn, which would have formed the basis of social reproduction—providing food, wares and crafts—and regional trade. The rapid and undemocratic termination of the land tenure system, and destruction of the economic system surrounding it, was, as the Zapatistas declared, a premeditated "death sentence" for the rural Indigenous communities whose livelihoods depended upon these lands (commons) and economic exchanges.

We can think of no greater act of re-commoning in this context, than of reconstituting *anew* traditional forms of social and economic organization; we add "anew" to highlight the spirit of intentional regeneration at the heart of communities like Tonatierra. As Reyes-Aguirre stated in response to my comment about contemporary communities reaching into the past for solutions to contemporary problems, she stated, "we reach into the past, yes, but *for what works.*" While contemporary calpolli are symbolic (prefigurative) in the sense that they have no official recognition or power, they are also highly functional insomuch as they bring people together in forms of ancestral and elected kinship that foster social support and coordination. They *work* because they are built on bonds of kinship that are based in trust and accountability and are reaffirmed regularly through ceremony and practice. Though no form of human social organization is perfect, forms like the calpolli provide the social and ethical grounds necessary for the vulnerable work of economic exchange and political activism.

While Tonatierra is not a "community economy" in the straightforward sense, we believe it is evocative of a more expansive and radical view of the economy. Indeed, as an organization dedicated to the regeneration of Indigenous communities, reproduction takes center stage in the organization's day-to-day activities and in the identities of members; and production finds its "traditional" place as a means to an end. In addition to being a reproductive commons (of knowledge, tradition, care and justice), Tonatierra is also a medium through which "other-than-capitalist subjectivities" are being prefigured and cultivated through ritualized practices and interactions (Gibson-Graham, 2006). While members of the community necessarily and actively exist and engage in the world outside Tonatierra, being a member of Tonatierra is an opportunity to exist and interact otherwise. One illustrative example involves the intentional ways in which the community articulates and translates the terminological requirements of the "non-profit world" into

Indigenous terms. So whereas the non-profit world requires listing a group of three "officers" including a "President," "Treasurer" and "Secretary," all internal documents (e.g. by-laws) are articulated in terms of a "Tenamaztle" consisting of a "QuetzalCoatl" (the person who represents the will of the community), the "Tezactlipoca" (the person who reflects and records the community) and the "Huitzilopochtli" (the person responsible for the offspring of the community). As Reyes-Aguirre explained, these positions (and the spiritual and traditional frameworks behind them) are what the children see and reproduce.

When we first learned about Tonatierra two years ago, we thought we were going to write about a coffee cooperative. As we hope this chapter has demonstrated, Tonatierra is so much more. This is not to underestimate the significance of the Quetzal Co-op; it is essential to the community's symbolism, formation and reproduction. In addition to being the thread that continues to weave north–south relations, it is also the community's historical through line, threading past, present and future. In addition to the Quetzal Co-op, this chapter has also focused on the "calpolli" as a symbolic and strategic form that allows Tonatierra to imagine other ways of being, relating, organizing and doing the difficult work of healing what Reyes-Aguirre described as the "medicine lines."

Notes

1 This quotation comes from the organization's webpage and is credited to founding member Tupac Enrique Acosta.

2 Also spelled calpulli.

3 Nahuatl is the language of the Nahua (Aztec) people that was spoken mainly in Central Mexico before Spanish colonization; it is still spoken today in parts of Mexico by more than 1.5 million people (see Mexikatlahtolli/Nawatlahtolli, 2018).

4 Hardt and Negri (and others) assert that "the development of neoliberal globalization from the 1970s was really a response to the 1960s confluence or accumulation of worker rebellions, liberation struggles and revolutionary movements throughout the world" (2019: 76); see also Arrighi 1994 and Jamieson 1997.

5 Chapter 2 provides a brief description of the forces of, and reactions to, neoliberalism in Latin America.

6 In 2010 they were a plaintiff in the ACLU federal lawsuit against AZ SB 1070; they represent Indigenous interests at the UN, and more recently they have been involved in projects helping undocumented migrants and migrant workers.

7 In the plural sense inspired by Gloria Anzaldúa (e.g. 1987, 2002, 2015).

8 There is evidence to suggest as well that the southwestern United States was the ancient homeland of the Aztecs (see http://americashistorias.blogspot.com/2015/)

9 Above the Gila river.

10 Below the Gila river.

11 From 1861 to 1862, the southern part of Arizona (below the Gila river) was a part of the Confederacy.

12 Poem used with permission of the author.

13 The Braceros program was one in which the legal status of the laborer was tied directly to the employer. If a worker lost favor with his employer, the worker might be plunged into "illegal alien" status. While some employers did not engage in abusive behaviors, a system designed as such has abuse and submission built in.

14 Internal critiques of this period of activism problematized the movement's romanticization of ancient Indigenous pasts that revealed "considerable ignorance about the realities of present-day Indigenous communities throughout the Americas" (MacCaughgan, 2020, p. 17).

15 Mexico joined GATT in 1986.

16 Lyrics from the (2017) song, "You build a wall" by British protest singer, song writer, Grace Petrie on the album Heart First Aid Kit. Lyric used with permission of the author.

17 While regional Indigenous movements addressing Indigenous rights in politics, education and labor developed in some parts of Mexico in the 1970s, the culture of resistance in Mexico was greatly influenced by the long rule (1929–2000) of the PRI—a party that absorbed internal distinctions around their dominant ideology of "revolutionary nationalism" and attempted to assimilate Indigenous peoples into a "mestizo national culture" in efforts that often involved the violent suppression of political activism (Grammont, Mackinlay and Stoller 2009, p. 23, 27). This was compounded by the bureaucratic inefficiency of the Nacional Campesina (CNC), the party formed in 1938 to represent rural (and mostly Indigenous) populations. See Bretón et al. (2021) for further discussion on the resistance led by a number of farmer and peasant organizations in 2003 after Mexico liberalized a number of agricultural commodities.

18 While the ILO has a checkered history in its concerns around "native workers," it has consistently been the first major international organization to get involved in "native" or "indigenous" issues (Niezen, 2003).

19 This is the first line from poet Alberto Ríos's (2014) poem, The border: A double sonnet.

20 Tonatierra works with various Indigenous communities to sell their wares in response to community needs.

21 This is a quotation from the Tonatierra website (see Danza Huehuecoyotl, 2022).

22 The Zapatistas were one of many other Indigenous groups organizing at this time.

References

Bretón, V., González, M., Rubio, B. and Vergara-Camus, L. (2022) Peasant and indigenous autonomy before and after the Pink tide in Latina America. *Journal of Agrarian Change.* doi: 10.1111/joac.12483

Buchanan, R. (1995) Border crossings: NAFTA, regulatory restructuring and the politics of place. *Indian Journal of Global Legal Studies*, 2(2), pp. 371–393.

Flores-Macías, G. A. and Sánches-Talanquer, M. (2019) The political economy of NAFTA/USMCA. *Oxford Encyclopedia of Politics*. Oxford: Oxford University Press.

Gibson-Graham, J.K. (2006). *A postcapitalist politics*. Minneapolis: University of Minnesota Press.

Godelmann, R.I. (2014) The Zapatista Movement: The fight for Indigenous rights in Mexico. *Australian Institute of International Affairs.* Available at: www.internationalaffairs.org.au/news-item/the-zapatista-movement-the-fight-for-indigenous-rights-in-mexico/

Grammont, de. H., Mackinlay, H. and Stoller, R. (2009) Campesino and Indigenous social organizations facing democratic transition in Mexico, 1938–2006. *Latin American Perspectives, Peasant Movements in Latin America: Looking Back, Move Ahead*, 36(4), pp. 21–40.

Haddal, C.C. (2010). *People crossing borders: An analysis of U.S. border protection policies*. Congressional Research Service, Washington DC. Retrieved from: https://sgp.fas.org/crs/homesec/R41237.pdf

Harvey, D. (2005). A Brief History of Neoliberalism. Oxford: Oxford University.

Hesketh, C. (2016) The survival of non-capitalism. Environment and Planning D: Society and Space, 34(5), pp. 877–894.

Hardt, M. and Negri, A. (2019) Empire, twenty years one. *New Left Review*, 120, pp. 67–92.

Harvey, N. (1995) Rebellion in Chiapas: Rural reforms and popular struggle. *Third World Quarterly*, 16(1), pp. 39–73.

Hesketh, C. (2016) The survival of non-capitalism. *Society and Space*, 34(5), pp. 877–894.

Huehuecoyotl (2022) Tonatierra. Retrieved on May 16, 2022 from www.tonatierra.org/huehuecoyotl

McAtackney, L and McGuire, R. H. (2020). *Walling and walling out: Why are we building new barriers to divide us?* Albuquerque: University of New Mexico Press.

McGuire R.H. and Van Dyke, R.M. (2019) Crossing *la Línea*: Bodily Encounters with the U.S.- México Border in Ambos Nogales. Sheridan, T. E. and McGuire, R. H., The border and its bodies: The embodiment of risk along the US–México Line. Arizona: The University of Arizona Press, pp. 41–70.

Mexikatlahtolli/Nawatlahtolli (náhuatl) (2018) Secretaría de Cultura/Sistema de Información Cultural. Retrieved on June 27, 2022 from https://sic.cultura.gob.mx/ficha.php?table=frpintangible&table_id=701INEGI

McCaughan, E.J. (2020) "We didn't cross the border, The border crossed us": Artists' images of the US–Mexico border and immigration. *Latin and Latinx Visual Culture*, 2(1), pp. 6–31.

Naylor, L. (2018) Fair trade coffee exchanges and community economies. *Environment and Planning A: Economy and Space*, 50(5), pp. 1027–1046.

Niezen, R. (2003). The origins of the International Movement of Indigenous Peoples. In R. Niezen, *The origins of Indigenism: Human rights and the politics of identity*. Berkley: University of California Press. pp. 29–52.

North, D. (2013) We should remember the bracero program … and shudder. *Center for Immigration Studies.* Retrieved on June 20, 2022 from https://cis.org/North/We-Should-Remember-Bracero-Program-and Shudder#:~:text=Farm%20labor%20wages%20stagnated%20at,routinely%20underpaid%20and%20badly%20treated

Quetzal Co-op~Cafetzin (2022) Tonatierra. Retrieved on May 2, 2022 from www.tonatierra.org/quetzalcoop

Reyes-Aguirre (2021, July 15) Interview.

Reyes-Aguirre (2021, October 4) Interview.

Reyes-Aguirre (2022, June 7) Interview.

Reyes-Aguirre (2022, July 9) Interview.

Ríos, A. (2014) The border: A double sonnet.

Ríos, A. (2007) Líneas fronterizas/border lines. *VQR*, 83(2).

Sheridan, T.E. (2012) *Arizona: A history (revised edition)*. Tucson: Arizona University Press.

Sheridan, T.E. and McGuire, R.H. (2019) *The border and its bodies: The embodiment of risk along the US–México Line*. Arizona: The University of Arizona Press.

Solano, L.X. (2007) "Indigenismo, indianissmo and 'ethnic citizenship' in Chiapas." *The Journal of Peasant Studies*, 32(3), pp. 555–583.

Tonatierra (2022) Huehuecoyotl. Retrieved on May 16, 2022 from www.tonatierra.org/huehuecoyotl

Tonatierra (2022a) Community~calpolli. Retrieved May 1, 2020 from www.tonatierra.org/community-calpolli

Tonatierra (2022b) Mission~vision. Retrieved on March 4, 2022 from www.tonatierra.org/mission

Tonatierra (2022) Nahuacalli Embassy of Indigenous Peoples. Retrieved on May 2, 2022 from www.tonatierra.org/neoip

United Nations (2007) United Nations Declaration on the Rights of Indigenous People. Retrieved from www.un.org/development/desa/indigenouspeoples/declaration-on-the-rights-of-indigenous-peoples.html

United Nations (1981) *International NGO Conference on Indigenous Peoples and the Land*. Geneva: Women's International League for Peace and Freedom.

Verea, M. (2014) Immigration trends after 20 years of NAFTA. *Norteamérica*, 9(2), pp. 109–143.

7 Conclusion

While this book project was conceived of in the year preceding the Covid crisis, the pandemic's initial impact and continuing aftershocks have continued to inform the book. In addition to bringing the world to a standstill—disrupting global supply chains and destroying lives and livelihoods—the pandemic also drew attention to the place of labor in our societies and lives. Whereas some people temporarily worked from home and others lost their jobs, many workers were too "essential" to cut or keep safe at home. And while there was a fleeting moment of appreciation for "essential workers," this attention ebbed as news cycles shifted. Contemporary conversations on labor have turned on the one hand to the standard topics of "labor shortages" ("people just don't want to work …") and "job growth" and on the other hand to a new topic, the so-called "Great Resignation." While none of these frames adequately represents or addresses the crisis of work in our times, they do indicate a seismic and long overdue shift in how we think about labor. In addition to highlighting the problems of "production" in our contemporary global capitalist system, the COVID crisis has also drawn attention to the pervasive crisis of "reproduction" in neoliberalism: a crisis long in the making. The crisis of reproduction is the result of more than 40 years of neoliberal policies that sold "FREEDOM!!!" but delivered enclosure. As noted throughout the book, the rhetoric and practices of neoliberalism must be understood in the broader context of the postwar period, most significantly the processes of decolonization and related civil rights movements of that period.

Our book approached the crisis of capitalism from the lens of "social imaginaries" situating capitalism not as an institution or system but rather as a hegemonic social formation that has come to dominate how we think about what the economy is and should be and has had very material implications for who and what we value. From the perspective of the imagination, the economy is a product not of history or of the will of a few powerful people but rather the product of the collective and creative of will of the people (and their histories and material realities). Guided by Haiven and Khasnabish's (2014) approach to the "radical imagination" and Gibson-Graham's (2006)

DOI: 10.4324/9781003138617-7

"community economies," our project examined three communities that are imagining economic relations and subjectivities "otherwise." Each of the three sites studied—Tombreck, Navdanya and Tonatierra—are communities of creative resistance that are responding to the colonial and neoliberal "crisis of reproduction" by engaging in mundane and everyday acts of imagining and laboring otherwise. Each of these communities engages in practices of re-commoning that challenge neoliberal and neocolonial practices of enclosure. Each case is also a provocative case of imagining labor not as the domain of capital or as limited to "productive" (waged) work but as a communal resource and activity whose primary purpose is to serve the community's reproduction. Because we could not do site visits, we were not able to spend the time we would have liked to learn about the limitations, hardships and failures endured by each community in the project of social reproduction. We understand that these hardships are an essential but largely undocumented feature of reproductive labor. In the same vein, we would like to follow up with research that addresses the incredible amount of invisible labor that goes into social reproduction of the sort taken on by the communities we studied.

This project was in part inspired by an attempt to find "hope in the dark" (Solnit, 2004) *not* in a powerful social or political movement or in a new approach to development but in the small and mundane acts happening all-the-time and everywhere around us. Each of these communities has filled our hope tanks in surplus. This is not because the organizations are perfect examples of utopias-in-the-making. It is because they have changed how we think about social change, activism and world building. As stated in Chapter 3, this book is the beginning of what we hope is a much longer project with each community. We hope that the coming years find us able to continue this research with individual site visits and in-depth interviews. We are especially keen to develop our approach to "solidarity research" such that our methods, findings and output are better informed by each community's needs and perspectives. We are enthusiastic as well to explore different "outcomes" for our research that might better serve the needs of economic reimagination and subvert typical academic output metrics. We are keen to spend more time with each site; studying them as distinct "community economies" responsive to unique histories, resources and needs.

As we stated in the introduction, writing a book about the problems and possibilities of labor while laboring in an economic sector that is both rapidly changing and also being exposed for its various abuses of labor, against the backdrop of a world economy in flux and a global population increasingly aware of the reproductive crisis in which we find ourselves, has been, well, very "meta." We are looking forward to stepping back a bit, reading the research that has been produced alongside ours during this "unprecedented" moment, reflecting, recharging and returning to these sites and this topic in hopes of keeping the flame of economic re-imagination ignited. We hope others will be inspired to do so as well.

References

Gibson-Graham, J.K. (2006) *The end of capitalism (as we knew it): A feminist critique of political economy*. Minneapolis: University of Minnesota Press.

Haiven, M. and Khasnabish, D.A. (2014) *The radical imagination: Social movement research in the age of austerity*. London: Bloomsbury Publishing.

Solnit, R. (2004) *Hope in the dark: Untold histories, wild possibilities*. Chicago: Haymarket.

Index

Note: Figures in this index are shown in *italics*. Endnotes consist of the page number plus 'n' and the endnote number i.e., 69n9 refers to endnote 9 on page 69.

For Product Safety Concerns and Information please contact our EU
representative GPSR@taylorandfrancis.com
Taylor & Francis Verlag GmbH, Kaufingerstraße 24, 80331 München, Germany

www.ingramcontent.com/pod-product-compliance
Ingram Content Group UK Ltd.
Pitfield, Milton Keynes, MK11 3LW, UK
UKHW020000200926
475702UK00015B/2391
* 9 7 8 0 3 6 7 6 8 6 8 6 4 *